RSC Paperbacks

# THE MISUSE OF DRUGS ACT
## A Guide for Forensic Scientists

L.A. KING

*Forensic Science Service, London Laboratory*
*London, UK*

# RS•C
advancing the chemical sciences

ISBN 0-85404-625-9

A catalogue record for this book is available from the British Library

Published by The Royal Society of Chemistry,
Thomas Graham House, Science Park, Milton Road,
Cambridge CB4 0WF, UK
Registered Charity Number 207890

For further information see our web site at www.rsc.org.

Typeset in Great Britain by Alden Bookset, Exeter, Devon, UK
Printed in Great Britain by TJ International, Padstow, Cornwall

# Preface

The primary role of the forensic drug chemist is to examine items submitted by law enforcement agencies and to determine if controlled (scheduled) substances are present. If that process required no more than a comparison of analytical results with a table of controlled drugs then there would be little need for this book. In reality, the forensic scientist is also required to place analytical findings in a wider context and to offer expert opinion in a court of law. All legislation is liable to interpretation; and the Misuse of Drugs Act 1971 is no exception. Indeed, because of its technical nature and the high political profile of drug abuse, this Act has undergone more scrutiny than most.

As an example of technical complexity in the Act, consider generic definitions. Some examples can be found in the international drug conventions, but the UK legislation takes them to a sophisticated level. The well-known controlled drug MDMA, a member of the 'Ecstasy' group of so-called entactogenic stimulants, is more formally known as 3,4-methylenedioxymethylamphetamine or fully systematically as *N*-methyl-1-(3,4-methylenedioxyphenyl)propan-2-amine. Yet none of these terms nor any other direct synonym will be found in the Act. One will likewise search in vain for 'heroin', 'LSD' or 'THC'. Alternatively, what is one to make of the following name: 2-(5-methoxy-2,2-dimethyl-2,3-dihydrobenzo-[*b*]furan-6-yl)-1-methylethylamine? Even an experienced chemist might struggle to recognise this immediately as a relative of MDMA. Yet that name and over thirty more like it were added to the Act in 2001. In this book, these 'new' phenethylamines, which are not covered by the current generic definitions, are described in some detail. Beyond the Act, a further level of complexity is introduced by the Misuse of Drugs Regulations 2001, which modify the provisions of the Act in certain situations.

This book is aimed not only at forensic scientists but also at police and customs officers, lawyers and all those who come into contact with the Misuse of Drugs Act. It is intended to clarify the many scientific complexities in the Act. For chemists, the extensive use of molecular structures in the text allows a complete and easier comprehension of the chemical background to the legislation, but they are not essential for a broad understanding. It is not a guide to general aspects of the law, stated cases, sentencing policy or related legislation, although some of these are dealt with briefly in Appendices. Also excluded is any general discussion of chemical analysis, but specific problem areas are briefly mentioned where they have a bearing on interpretation. No account is provided of the social dimension to drug abuse, epidemiology, pharmacology or toxicology, but the interested reader is directed to the Bibliography. A number of possible changes to the Misuse of Drugs Act and the Misuse of Drugs Regulations are included in Chapter 9. These future developments are largely concerned with technical issues, although the recent suggestion that cannabis and cannabis resin might be re-classified could have major implications for the administration of justice. This guide does not include any strategic view of drug control; there is no discussion of the arguments for or against legalisation or decriminalisation of some or all drugs. Finally, it is beyond the scope of this book to provide any recommendations on the presentation of evidence in court or how analytical results should be set out in reports and statements. Although intended as a guide to the United Kingdom law, many of the provisions of the Misuse of Drugs Act originate in international treaties and have a wider relevance.

I am particularly grateful for the support of the Secretariat of the Home Office Advisory Council on the Misuse of Drugs. Together with fellow members of the Council and its subcommittees, they have provided much stimulating discussion, resolved numerous problems and illuminated the darker corners of the Misuse of Drugs Act. Dr Mike Griffin and Professor Geoffrey Phillips kindly reviewed a draft manu-script and offered valuable comments. I would also like to thank many former colleagues in the Forensic Science Service and other laboratories, members of the legal profession and law enforcement officers for their challenging questions and helpful comments over many years.

This book is dedicated to my wife, Diane.

*Leslie A. King*
*Hampshire*
*April, 2002*

# Contents

*Chapter 1*

# Introduction

*Possession is not nine-tenths of the law – it's nine-tenths of the problem.*

John Lennon

## 1.1  DRUG ABUSE

Drugs, whose possession or supply is restricted by law, are known as controlled or scheduled substances. They comprise both licit materials (*i.e.* those manufactured under licence for therapeutic use) and the illicit products of clandestine factories. Although many plant-based drugs have been self-administered for thousands of years (*e.g.* coca leaf, cannabis, opium, peyote cactus), the imposition of criminal sanctions is mostly a product of the 20th century. Many of the drugs currently abused were once not only on open sale, but often promoted as beneficial substances by the food and pharmaceutical industries. A pattern developed whereby initial misuse of pharmaceutical products such as morphine, cocaine and amphetamine led to increasing legal restrictions and the consequent rise of an illicit industry. Nowadays, nearly all serious drug abuse involves illicitly-produced substances. Most fall into just a few pharmacological groups, *e.g.* central nervous system stimulants, narcotic analgesics, hallucinogens (psychotomimetics) and hypnotics. It is still true that the most prevalent drugs are the plant-derived or semi-synthetic substances (*e.g.* cannabis, cocaine and heroin), but the view of the United Nations Drug Control Programme is that wholly synthetic drugs (*e.g.* amphetamine, MDMA and related designer drugs) are likely to pose a more significant social problem in the future. According to the World Health Organisation (WHO), scheduled drugs are 'abused' rather than 'misused', but in the following text the two terms are used synonymously. Drugs of abuse may or may not lead to

1

physical or psychological 'dependence': a term used by WHO in preference to 'addiction'.

On the basis of a recent Home Office survey (Drug Misuse Declared in 2000 – see Bibliography), a third of the adult population in the United Kingdom (UK) admits to having used a controlled drug at least once in their lives; fewer than 10% use drugs on a regular basis and for the great majority of these the drug involved is cannabis. The next most common drugs are amphetamine, cocaine and 3,4-methylenedioxymethylamphetamine (MDMA). Seizure data from police and customs show a broadly similar pattern. There are currently over 100 000 arrests each year in the UK for drug offences, again the majority involving cannabis. In Europe, it is estimated that 0.2–0.3% of the population are regular heroin users. With few exceptions, the scale of drug abuse has steadily increased in most countries, but it is still predominantly associated with younger members of the population. Mortality from drug abuse has risen rapidly and is largely associated with opiates.

## 1.2   UK DRUG CONTROL LEGISLATION BEFORE 1971

Apart from the Pharmacy Act of 1868, which restricted the sale of opium, the modern period of drug control started in the early 20th century. Following the Poisons and Pharmacy Act 1908 and the Shanghai Opium Commission in 1909, further restrictions were introduced on cocaine, morphine and opium. More controls on a wider range of substances were introduced by the successive Dangerous Drugs Acts of 1920, 1925, 1951 and 1964. Synthetic amphetamine-like drugs entered the legislation in the Drugs (Prevention of Misuse) Act (DPMA) 1964, and a Modification Order in 1966 added lysergide (LSD) to the DPMA. The Dangerous Drugs Act 1965 consolidated previous legislation.

The first attempts to introduce structure-related generic control into UK drugs law were made with the DPMA of 1964. This contained a statement intended to cover a range of synthetic stimulants. The key feature was a definition of certain side-chain substitution patterns in α-methylphenethylamine (amphetamine) and β-methylphenethylamine. While this did indeed encompass compounds such as phentermine, methylphenidate and diethylpropion, it soon became clear that a refined interpretation of the generic statement unwittingly included dozens of drugs that were not stimulants. In fact, it could be argued that some barbiturates were also covered. This generic control was repealed by a Modification Order in 1970. Following this early failure, it would be some years before generic control of phenethylamines again entered the legislation. But this time (1977), the focus would be on ring-substituted phenethylamines, it would be much more robust and would be followed by generic controls for several other groups.

## 1.3  THE UNITED NATIONS CONVENTIONS

In international law, controls on drugs of abuse are set out in three United Nations (UN) treaties: The Single Convention on Narcotic Drugs 1961 (UN1961), the Convention on Psychotropic Substances 1971 (UN1971) and The Convention Against Illicit Traffic in Narcotic Drugs and Psychotropic Substances 1988 (UN1988). These treaties are implemented in domestic laws by signatory states, and have been considerably extended in some. In the UK, the corresponding legislation is the Misuse of Drugs Act 1971. Since the inception of the UN conventions, numerous substances have been added to the Schedules, particularly those of the 1971 treaty.

In the 1961 convention, there is a strong emphasis on plant-based drugs (*i.e.* cannabis, opium and cocaine), with rules for their cultivation, manufacture and trade. In addition, over 100 other substances, mostly synthetic narcotic analgesics, are included, but only a few of these are now used clinically or ever abused. More than 100 psychotropic substances are listed in the 1971 convention, but again only a small fraction is regularly abused. Unlike the treaty of 1961, there is no overarching control of the stereoisomers of psychotropic drugs. Thus in Schedule I, amphetamine, meaning both the (−) and the (+) enantiomers, is listed together with dexamphetamine [the (+) enantiomer of amphetamine] and levamphetamine [the (−) enantiomer] while methamphetamine, meaning the (+) enantiomer, is listed alongside methamphetamine racemate [a mixture of the (−) and (+) enantiomers]. These examples, and the situation whereby the stereochemical configuration of many other substances was left unspecified, has lead to some confusion, but the UN has recently moved to rationalise this matter. These problems have been avoided in the Misuse of Drugs Act by the inclusion of the stereoisomers of almost all controlled drugs (see Chapter 3).

## 1.4  PRINCIPLES OF CURRENT UK LEGISLATION

The Misuse of Drugs Act 1971 replaced The Dangerous Drugs Act 1965 and the DPMA 1964 and introduced the concept of 'controlled drugs'. These are defined as those substances or products set out in Schedule 2. The Misuse of Drugs Act, which came into effect in 1973, set up an 'Advisory Council on the Misuse of Drugs' whose terms of reference include a statement of what might constitute a controlled drug. This is set out in Section 1(2) of the Act as: *'It shall be the duty of the Advisory Council to keep under review the situation in the United Kingdom with respect to drugs which are being or appear to them likely to be misused and of which the misuse is having or appears to them capable of having harmful effects sufficient to constitute a social problem. …'.*

The Misuse of Drugs Act prohibits certain activities with respect to controlled drugs (*e.g.* possession, possession with intent to supply, production) without a licence. With the exception of opium, there is no illegality in using or consuming a controlled drug. The great majority of arrests for offences under the Misuse of Drugs Act involve possession of relatively small amounts of a controlled drug. The drugs are listed in Schedule 2 of the Act and are divided into three Classes: Class A (Part I of Schedule 2), Class B (Part II) and Class C (Part III). These classes represent, in decreasing order A to C, the propensity of the substances to cause social harm. Associated with each class are the maximum penalties for offences involving controlled drugs, again decreasing in severity in the order A to C. For Class A drugs, the maximum penalty for some offences is life imprisonment, for Class B it is 14 years and for Class C, five years. A tariff of penalties has been established by the courts for a range of situations and some of these are described in Appendix 8. The list of substances in Schedule 2 may be varied by a Statutory Instrument known as a Modification Order. There have been 14 such Orders since 1971 (see Appendix 1), most of which have served to incorporate changes agreed by member states of the UN.

The details of control are legislated by 'regulations'. In the Misuse of Drugs Regulations 2001, which came into force on 1st February 2002 and replaced the previous Regulations of 1985, controlled drugs are divided into five schedules. The organisational principles of the Schedules in the Regulations are shown in Appendix 3. In simple terms, the Regulations set out what *should* be done with controlled drugs (*i.e.* their licit use) whereas the Act sets out what *should not* be done (*i.e.* their illicit use).

Many of the substances listed in the Misuse of Drugs Act and the Misuse of Drugs Regulations, numerous definitions and some parts of the generic controls derive directly from the UN conventions. However, the Misuse of Drugs Act goes beyond the minimum in several important areas. Not only are there more substances, but an important feature of the Act is the extensive use of structure-related generic terms. Because the Misuse of Drugs Act relies on the concept of actual or potential social harm, rather than the specific pharmacological or toxicological properties of a controlled drug, no great difficulty arises from the introduction of generic control. It is quite certain that amongst the essentially infinite number of generically-defined substances there will be compounds that have no abuse potential and some may have no physiological effect of any sort. Without these effects, a substance will not be marketed by the pharmaceutical industry and neither will it be produced as a misusable drug. The structure-related definitions discussed in Chapter 5 are found only in the Misuse of Drugs Act (and the Regulations) and closely-related legislation such as the Misuse of Drugs Act of the Republic of Ireland 1976, although similar concepts, differing only in detail, occur in the drugs legislation of New Zealand and parts of Australia.

Nearly all substances in the Act are chemically-defined entities. Although a number of controlled drugs are found in a variety of plants, the only 'vegetable' materials specifically controlled are coca leaf, opium, poppy-straw and its concentrate (all Class A) and cannabis and cannabis resin (both Class B, but see Chapter 9). A separate offence of cultivation also exists for these materials, although it is now common for cultivation to be subsumed under the broader offence of production of a controlled drug. There is a general reluctance to control too many botanical entities, partly because of taxonomic difficulties and partly because they may be seen as ubiquitous and naturally occurring.

Appendix 9 provides brief details of other legislation concerned with drug control. Apart from the Misuse of Drugs Act, the forensic scientist will have occasional need to consult the Medicines Act, 1968, perhaps to determine if a submitted drug is a prescription only medicine. However, the definition of a medicinal product in that Act has now been super-seded (see Chapter 8). There are legal restrictions on the trade in certain substances used in the manufacture of drugs (so-called precursor chemicals), brief details of which are shown in Appendix 6.

## 1.5   EUROPEAN INITIATIVES

In June 1997, the Council of the European Union (EU) adopted a so-called 'Joint Action' on new synthetic drugs (NSD). This was concerned with information exchange on new substances as well as procedures for risk assessment and legal control. 'New synthetic drugs' were defined as those substances that had little or no therapeutic value and were not already under international control, but had a potential for abuse similar to those substances listed in Schedules I and II of the United Nations Convention on Psychotropic Substances 1971. The basis for this agreement was Article K.3 of the Treaty on European Union, signed at Maastricht, concerning the approximation of the laws and practices of the Member States of the European Union to combat drug addiction and to prevent and combat illegal drug trafficking. The Joint Action was promoted by the Dutch Government and it took place against a political background whereby Europe had become a leading producer of these drugs. Difficulties were emerging in the EU because of the way in which certain drugs were treated differently in different Member States.

The Joint Action was intended to focus attention on the plethora of 'designer drugs' (nearly all phenethylamines), which had appeared during the 1990s. A typical feature of NSD is that they are often presented in the form of white tablets bearing 'illicit' logos, which provide no clue to the chemical contents. Well-established drugs such as amphetamine, MDMA and its ethyl analogue (MDEA) were excluded since they were already controlled in international law.

Since 1997, of the several NSD reported in the EU, three (*N*-methyl-1-(1,3-benzodioxol-5-yl)-2-butanamine [MBDB], 4-methylthioamphetamine [4-MTA] and *para*methoxymethylamphetamine [PMMA]) were selected for risk assessment by the European Monitoring Centre for Drugs and Drug Addiction (EMCDDA). The main elements of the risk assessment criteria were a chemical and pharmacological description of the NSD, acute and chronic psychological risks, sociological and criminological evidence, and public health risks. A recommendation was subsequently made that 4-MTA (4-methylthioamphetamine) should become a scheduled drug in all Member States. Although neither gamma-hydroxybutyrate (GHB) nor ketamine fall within the definition of an NSD, they were subjected to risk assessment by EMCDDA in 2000 because both are widely abused and neither is controlled under UN1971. No clear evidence emerged that either GHB or ketamine should be recommended for control in the EU, but the decision on GHB was overtaken by a meeting of the United Nations Commission on Narcotic Drugs (UN CND) in early 2001 where it was agreed that GHB should be brought under Schedule IV of UN1971 (see Chapter 9).

## 1.6  ABBREVIATIONS

In some publications, the abbreviation MDA is used for the Misuse of Drugs Act 1971, but this is not ideal because MDA is also the common acronym for 3,4-methylenedioxyamphetamine: one of the 'Ecstasy' drugs. In this book, the Misuse of Drugs Act 1971 is referred to as the 'Act'. The Misuse of Drugs Regulations, 2001 are shown as the 'Regulations'. Substances listed in Schedule 2 to the Act are correctly known as 'controlled drugs', but, for the sake of clarity and when the context is clear, are often described herein simply as 'drugs' or 'substances'. By normal convention, the abbreviation 'mg' is used in the following text for milligram(s). It may be noted that, until the 2001 revision, the Regulations used the obsolete term milligrammes.

*Chapter 2*

# Schedule 2 to the Act

The following tables and subordinate text are set out with paragraph headings as they appear in Schedule 2 to the Act. Class A controlled drugs are included in Part I of Schedule 2, Class B in Part II and Class C drugs in Part III. Substances or products, which are listed in UN1961 or UN1971, are indicated along with the corresponding Schedule of the Regulations (see also Appendices 3 to 5). Where a substance or generic definition was introduced into the Act by a subsequent Modification Order (see Appendix 1), the corresponding Statutory Instrument Number (S.I.) and date are shown.

**Table 2.1** *Class A controlled drugs listed in Part I of Schedule 2 to the Act*

1. The following substances and products, namely:

(a)

| Substance or Product | UN Convention and Schedule | Modification Order | Schedule in Regulations |
|---|---|---|---|
| Acetorphine | UN1961 (I) | | 2 |
| Alfentanil | UN1961 (I) | (S.I. 859)1984 | 2 |
| Allylprodine | UN1961 (I) | | 2 |
| Alphacetylmethadol | UN1961 (I) | | 2 |
| Alphameprodine | UN1961 (I) | | 2 |
| Alphamethadol | UN1961 (I) | | 2 |
| Alphaprodine | UN1961 (I) | | 2 |
| Anileridine | UN1961 (I) | | 2 |
| Benzethidine | UN1961 (I) | | 2 |
| Benzylmorphine (3-benzylmorphine) | UN1961 (I) | | 2 |
| Betacetylmethadol | UN1961 (I) | | 2 |
| Betameprodine | UN1961 (I) | | 2 |

*(Continued)*

**Table 2.1**  *Continued*

| Substance or Product | UN Convention and Schedule | Modification Order | Schedule in Regulations |
|---|---|---|---|
| Betamethadol | UN1961 (I) | | 2 |
| Betaprodine | UN1961 (I) | | 2 |
| Bezitramide | UN1961 (I) | | 2 |
| 4-Bromo-2,5-dimethoxy-α-methylphenethylamine | UN1971 (I) | (S.I. 421)1975 | 1 |
| Bufotenine | Not listed | | 1 |
| Cannabinol, except where contained in cannabis or cannabis resin | UN1971 (I) | | 1 |
| Cannabinol derivatives | UN1971 (I) (Dronabinol is in Schedule II) | | 1 (Dronabinol is in Schedule 2) |
| Carfentanil | Not listed | (S.I. 2230)1986 | 2 |
| Clonitazene | UN1961 (I) | | 2 |
| Coca leaf | UN1961 (I) | | 1 |
| Cocaine | UN1961 (I) | | 2 |
| 4-Cyano-2-dimethylamino-4,4-diphenylbutane | UN1961 (I) | | 2 |
| 4-Cyano-1-methyl-4-phenylpiperidine | UN1961 (I) | | 2 |
| Desomorphine | UN1961 (I) | | 2 |
| Dextromoramide | UN1961 (I) | | 2 |
| Diamorphine | UN1961 (I) | | 2 |
| Diampromide | UN1961 (I) | | 2 |
| Diethylthiambutene | UN1961 (I) | | 2 |
| *N,N*-Diethyltryptamine | UN1971 (I) | | 1 |
| Difenoxin [1-(3-cyano-3,3-diphenylpropyl)-4-phenylpiperidine-4-carboxylic acid] | UN1961 (I) | (S.I. 421)1975 | 2 |
| Dihydrocodeinone *O*-carboxymethyloxime | UN1961 (I) | | 2 |
| Dihydromorphine | UN1961 (I) | | 2 |
| Dimenoxadole | UN1961 (I) | | 2 |
| Dimepheptanol | UN1961 (I) | | 2 |
| Dimethylthiambutene | UN1961 (I) | | 2 |
| *N,N*-Dimethyltryptamine | UN1971 (I) | | 1 |
| 2,5-Dimethoxy-α,4-dimethylphenethylamine | UN1971 (I) | | 1 |
| Dioxaphetyl butyrate | UN1961 (I) | | 2 |
| Diphenoxylate | UN1961 (I) | | 2 |
| Dipipanone | UN1961 (I) | | 2 |
| Drotebanol (3,4-dimethoxy-17-methylmorphinan-6β,14-diol) | UN1961 (I) | (S.I. 771)1973 | 2 |

*(Continued)*

**Table 2.1** *Continued*

| Substance or Product | UN Convention and Schedule | Modification Order | Schedule in Regulations |
|---|---|---|---|
| Ecgonine, and any derivative of ecgonine which is convertible to ecgonine or to cocaine | UN1961 (I) | | 2 |
| Ethylmethylthiambutene | UN1961 (I) | | 2 |
| Eticyclidine | UN1971 (I) | (S.I. 859)1984 | 1 |
| Etonitazene | UN1961 (I) | | 2 |
| Etorphine | UN1961 (I) | | 2 |
| Etoxeridine | UN1961 (I) | | 2 |
| Etryptamine | UN1971 (I) | (S.I. 750)1998 | 1 |
| Fentanyl | UN1961 (I) | | 2 |
| Furethidine | UN1961 (I) | | 2 |
| Hydrocodone | UN1961 (I) | | 2 |
| Hydromorphinol | UN1961 (I) | | 2 |
| Hydromorphone | UN1961 (I) | | 2 |
| Hydroxypethidine | UN1961 (I) | | 2 |
| N-Hydroxy-tenamphetamine | UN1971 (I) | (S.I. 2589)1990 | 1 |
| Isomethadone | UN1961 (I) | | 2 |
| Ketobemidone | UN1961 (I) | | 2 |
| Levomethorphan | UN1961 (I) | | 2 |
| Levomoramide | UN1961 (I) | | 2 |
| Levophenacylmorphan | UN1961 (I) | | 2 |
| Levorphanol | UN1961 (I) | | 2 |
| Lofentanil | Not listed | (S.I. 2230)1986 | 2 |
| Lysergamide | Not listed | | 1 |
| Lysergide and other N-alkyl derivatives of lysergamide | UN1971 (I) | | 1 |
| Mescaline | UN1971 (I) | | 1 |
| Metazocine | UN1961 (I) | | 2 |
| Methadone | UN1961 (I) | | 2 |
| Methadyl acetate | UN1961 (I) | | 2 |
| 4-Methyl-aminorex | UN1971 (I) | (S.I. 2589)1990 | 1 |
| Methyldesorphine | UN1961 (I) | | 2 |
| Methyldihydromorphine (6-methyldihydromorphine) | UN1961 (I) | | 2 |
| 2-Methyl-3-morpholino-1,1-diphenylpropane-carboxylic acid | UN1961 (I) | | 2 |
| 1-Methyl-4-phenylpiperidine-4-carboxylic acid | UN1961 (I) | | 2 |
| Metopon | UN1961 (I) | | 2 |
| Morpheridine | UN1961 (I) | | 2 |
| Morphine | UN1961 (I) | | 2 |

*(Continued)*

**Table 2.1** *Continued*

| Substance or Product | UN Convention and Schedule | Modification Order | Schedule in Regulations |
|---|---|---|---|
| Morphine methobromide, morphine *N*-oxide and other pentavalent nitrogen morphine derivatives | UN1961 (I) | | 2 |
| Myrophine | UN1961 (I) | | 2 |
| Nicomorphine (3,6-dinicotinoylmorphine) | UN1961 (I) | | 2 |
| Noracymethadol | UN1961 (I) | | 2 |
| Norlevorphanol | UN1961 (I) | | 2 |
| Normethadone | UN1961 (I) | | 2 |
| Normorphine | UN1961 (I) | | 2 |
| Norpipanone | UN1961 (I) | | 2 |
| Opium, whether raw, prepared or medicinal | UN1961 (I) | | 1 (raw opium) 2 (medicinal opium) |
| Oxycodone | UN1961 (I) | | 2 |
| Oxymorphone | UN1961 (I) | | 2 |
| Pethidine | UN1961 (I) | | 2 |
| Phenadoxone | UN1961 (I) | | 2 |
| Phenampromide | UN1961 (I) | | 2 |
| Phenazocine | UN1961 (I) | | 2 |
| Phencyclidine | UN1971 (II) | (S.I. 299)1979 | 2 |
| Phenomorphan | UN1961 (I) | | 2 |
| Phenoperidine | UN1961 (I) | | 2 |
| 4-Phenylpiperidine-4-carboxylic acid ethyl ester | UN1961 (I) | | 2 |
| Piminodine | UN1961 (I) | | 2 |
| Piritramide | UN1961 (I) | | 2 |
| Poppy-straw and concentrate of poppy-straw | UN1961 (I) (as concentrate) | | 1 (as concentrate) |
| Proheptazine | UN1961 (I) | | 2 |
| Properidine (1-methyl-4-phenylpiperidine-4-carboxylic acid isopropyl ester) | UN1961 (I) | | 2 |
| Psilocin | UN1971 (I) | | 1 |
| Racemethorphan | UN1961 (I) | | 2 |
| Racemoramide | UN1961 (I) | | 2 |
| Racemorphan | UN1961 (I) | | 2 |
| Rolicyclidine | UN1971 (I) | (S.I. 859)1984 | 1 |
| Sufentanil | UN1961 (I) | (S.I. 765)1983 | 2 |
| Tenocyclidine | UN1971 (I) | (S.I. 859)1984 | 1 |
| Thebacon | UN1961 (I) | | 2 |
| Thebaine | UN1961 (I) | | 2 |
| Tilidate | UN1961 (I) | (S.I. 765)1983 | 2 |
| Trimeperidine | UN1961 (I) | | 2 |

(b)   Any compound (not being a compound for the time being specified in sub-paragraph (a) above) structurally derived from tryptamine or from a ring-hydroxy tryptamine by substitution at the nitrogen atom of the side-chain with one or more alkyl substituents but no other substituent; (S.I. 1243)1977.

(ba)   The following phenethylamine derivatives, namely:

[All are unlisted in UN Conventions; all are (S.I. 3932)2001 and Schedule 1 in the Regulations]

---

Allyl(α-methyl-3,4-methylenedioxyphenethyl)amine
2-Amino-1-(2,5-dimethoxy-4-methylphenyl)ethanol
2-Amino-1-(3,4-dimethoxyphenyl)ethanol
Benzyl(α-methyl-3,4-methylenedioxyphenethyl)amine
4-Bromo-$\beta$,2,5-trimethoxyphenethylamine
*N*-(4-*sec*-Butylthio-2,5-dimethoxyphenethyl)hydroxylamine
Cyclopropylmethyl(α-methyl-3,4-methylenedioxyphenethyl)amine
2-(4,7-Dimethoxy-2,3-dihydro-1*H*-indan-5-yl)ethylamine
2-(4,7-Dimethoxy-2,3-dihydro-1*H*-indan-5-yl)-1-methylethylamine
2-(2,5-Dimethoxy-4-methylphenyl)cyclopropylamine
2-(1,4-Dimethoxy-2-naphthyl)ethylamine
2-(1,4-Dimethoxy-2-naphthyl)-1-methylethylamine
*N*-(2,5-Dimethoxy-4-propylthiophenethyl)hydroxylamine
2-(1,4-Dimethoxy-5,6,7,8-tetrahydro-2-naphthyl)ethylamine
2-(1,4-Dimethoxy-5,6,7,8-tetrahydro-2-naphthyl)-1-methylethylamine
α,α-Dimethyl-3,4-methylenedioxyphenethylamine
α,α-Dimethyl-3,4-methylenedioxyphenethyl(methyl)amine
Dimethyl(α-methyl-3,4-methylenedioxyphenethyl)amine
*N*-(4-Ethylthio-2,5-dimethoxyphenethyl)hydroxylamine
4-Iodo-2,5-dimethoxy-α-methylphenethyl(dimethyl)amine
2-(1,4-Methano-5,8-dimethoxy-1,2,3,4-tetrahydro-6-naphthyl)ethylamine
2-(1,4-Methano-5,8-dimethoxy-1,2,3,4-tetrahydro-6-naphthyl)-1-methylethylamine
2-(5-Methoxy-2,2-dimethyl-2,3-dihydrobenzo[*b*]furan-6-yl)-1-methylethylamine
2-Methoxyethyl(α-methyl-3,4-methylenedioxyphenethyl)amine
2-(5-Methoxy-2-methyl-2,3-dihydrobenzo[*b*]furan-6-yl)-1-methylethylamine
$\beta$-Methoxy-3,4-methylenedioxyphenethylamine
1-(3,4-Methylenedioxybenzyl)butyl(ethyl)amine
1-(3,4-Methylenedioxybenzyl)butyl(methyl)amine
2-(α-Methyl-3,4-methylenedioxyphenethylamino)ethanol
α-Methyl-3,4-methylenedioxyphenethyl(prop-2-ynyl)amine
*N*-Methyl-*N*-(α-methyl-3,4-methylenedioxyphenethyl)hydroxylamine
*O*-Methyl-*N*-(α-methyl-3,4-methylenedioxyphenethyl)hydroxylamine
α-Methyl-4-(methylthio)phenethylamine
$\beta$,3,4,5-Tetramethoxyphenethylamine
$\beta$,2,5-Trimethoxy-4-methylphenethylamine

---

(c)   Any compound (not being methoxyphenamine or a compound for the time being specified in sub-paragraph (a) above) structurally derived from phenethylamine, an *N*-alkylphenethylamine, α-methylphenethylamine, an *N*-alkyl-α-methylphenethylamine, α-ethylphenethylamine, or an *N*-alkyl-α-ethylphenethylamine by substitution in the ring to any extent with alkyl, alkoxy, alkylenedioxy or halide substituents, whether or not further substituted in the ring by one or more other univalent substituents. [(S.I. 1243)1977]

(d) Any compound (not being a compound for the time being specified in sub-paragraph (a) above) structurally derived from fentanyl by modification in any of the following ways, that is to say:

(i) by replacement of the phenyl portion of the phenethyl group by any heteromonocycle whether or not further substituted in the heterocycle;

(ii) by substitution in the phenethyl group with alkyl, alkenyl, alkoxy, hydroxy, halogeno, haloalkyl, amino or nitro groups;

(iii) by substitution in the piperidine ring with alkyl or alkenyl groups;

(iv) by substitution in the aniline ring with alkyl, alkoxy, alkylenedioxy, halogeno or haloalkyl groups;

(v) by substitution at the 4-position of the piperidine ring with any alkoxycarbonyl or alkoxyalkyl or acyloxy group;

(vi) by replacement of the *N*-propionyl group by another acyl group [(S.I. 2230)1986].

(e) Any compound (not being a compound for the time being specified in sub-paragraph (a) above) structurally derived from pethidine by modification in any of the following ways, that is to say:

(i) by replacement of the 1-methyl group by an acyl, alkyl whether or not unsaturated, benzyl or phenethyl group, whether or not further substituted;

(ii) by substitution in the piperidine ring with alkyl or alkenyl groups or with a propano bridge, whether or not further substituted;

(iii) by substitution in the 4-phenyl ring with alkyl, alkoxy, aryloxy, halogeno or haloalkyl groups;

(iv) by replacement of the 4-ethoxycarbonyl by any other alkoxycarbonyl or any alkoxyalkyl or acyloxy group;

(v) by formation of an *N*-oxide or of a quaternary base [(S.I. 2230)1986].

2. Any stereoisomeric form of a substance for the time being specified in paragraph 1 not being dextromethorphan or dextrorphan.

3. Any ester or ether of a substance for the time being specified in paragraph 1 or 2, not being a substance for the time being specified in Part II of this Schedule [(S.I. 771)1973].

4. Any salt of a substance for the time being specified in any of paragraphs 1 to 3.

5. Any preparation or other product containing a substance or product for the time being specified in any of paragraphs 1 to 4.

6. Any preparation designed for administration by injection specified in any of paragraphs 1 to 3 of Part II of this Schedule.

**Table 2.2** *Class B controlled drugs listed in Part II of Schedule 2 to the Act*

1. The following substances and products, namely:

(a)

| Substance or Product | UN Convention and Schedule | Modification Order | Schedule in Regulations |
|---|---|---|---|
| Acetyldihydrocodeine | UN1961 (II) | | 2 |
| Amphetamine | UN1971 (II) | | 2 |
| Cannabis and cannabis resin | UN1961 (I) | | 1 |
| Codeine | UN1961 (II) | | 2 |
| Dihydrocodeine | UN1961 (II) | | 2 |
| Ethylmorphine (3-ethylmorphine) | UN1961 (II) | | 2 |
| Glutethimide | UN1971 (III) | (S.I. 1995)1985 | 2 |
| Lefetamine | UN1971 (IV) | (S.I. 1995)1985 | 2 |
| Mecloqualone | UN1971 (II) | (S.I. 859)1984 | 2 |
| Methaqualone | UN1971 (II) | (S.I. 859)1984 | 2 |
| Methcathinone | UN1971 (I) | (S.I. 750)1998 | 1 |
| Methylamphetamine | UN1971 (II) | | 2 |
| α-Methylphenethyl-hydroxylamine | Not listed | (S.I. 3932)2001 | 2 |
| Methylphenidate | UN1971 (II) | | 2 |
| Methylphenobarbitone | UN1971 (IV) | (S.I. 859)1984 | 3 |
| Nicocodine | UN1961 (II) | | 2 |
| Nicodicodine (6-nicotinoyl-dihydrocodeine) | UN1961 (II) | (S.I. 771)1973 | 2 |
| Norcodeine | UN1961 (II) | | 2 |
| Pentazocine | UN1971 (III) | (S.I. 1995)1985 | 3 |
| Phenmetrazine | UN1971 (II) | | 2 |
| Pholcodine | UN1961 (II) | | 2 |
| Propiram | UN1961 (II) | (S.I. 771)1973 | 2 |
| Zipeprol | UN1971 (II) | (S.I. 750) 1998 | 2 |

**Table 2.2** *Continued*

(b)

| Substance or Product | UN 1961 or 1971 Conventions | Modification Order | Schedule in Regulations |
|---|---|---|---|
| any 5,5-disubstituted barbituric acid | UN1971 (III or IV) (Quinal-barbitone is in Schedule II) | (S.I. 859)1984 | 3 (Quinal-barbitone is in Schedule 2) |

2. Any stereoisomeric form of a substance for the time being specified in paragraph 1 of this Part of this Schedule.
3. Any salt of a substance for the time being specified in paragraph 1 or 2 of this Part of this Schedule.
4. Any preparation or other product containing a substance or product for the time being specified in any of paragraphs 1 to 3 of this Part of this Schedule, not being a preparation falling within paragraph 6 of Part I of this Schedule.

**Table 2.3** *Class C controlled drugs listed in Part III of Schedule 2 to the Act*

1. The following substances and products, namely:

(a)

| Substance or Product | UN Convention and Schedule | Modification Order | Schedule in Regulations |
|---|---|---|---|
| Alprazolam | UN1971 (IV) | (S.I. 1995)1985 | 4 (Part I) |
| Aminorex | UN1971 (IV) | (S.I. 750)1998 | 4 (Part I) |
| Benzphetamine | UN1971 (IV) | | 3 |
| Bromazepam | UN1971 (IV) | (S.I. 1995)1985 | 4 (Part I) |
| Brotizolam | UN1971 (IV) | (S.I. 750)1998 | 4 (Part I) |
| Buprenorphine | UN1971 (III) | (S.I. 1340)1989 | 3 |
| Camazepam | UN1971 (IV) | (S.I. 1995)1985 | 4 (Part I) |
| Cathine | UN1971 (III) | (S.I. 2230)1986 | 3 |
| Cathinone | UN1971 (I) | (S.I. 2230)1986 | 1 |
| Chlordiazepoxide | UN1971 (IV) | (S.I. 1995)1985 | 4 (Part I) |
| Chlorphentermine | Not listed | | 3 |
| Clobazam | UN1971 (IV) | (S.I. 1995)1985 | 4 (Part I) |
| Clonazepam | UN1971 (IV) | (S.I. 1995)1985 | 4 (Part I) |
| Clorazepic acid | UN1971 (IV) | (S.I. 1995)1985 | 4 (Part I) |
| Clotiazepam | UN1971 (IV) | (S.I. 1995)1985 | 4 (Part I) |
| Cloxazolam | UN1971 (IV) | (S.I. 1995)1985 | 4 (Part I) |
| Delorazepam | UN1971 (IV) | (S.I. 1995)1985 | 4 (Part I) |

*(Continued)*

**Table 2.3** *Continued*

| Substance or Product | UN Convention and Schedule | Modification Order | Schedule in Regulations |
|---|---|---|---|
| Dextropropoxyphene | Not listed | (S.I. 765)1983 | 2 |
| Diazepam | UN1971 (IV) | (S.I. 1995)1985 | 4 (Part I) |
| Diethylpropion | UN1971 (IV) | (S.I. 859)1984 | 3 |
| Estazolam | UN1971 (IV) | (S.I. 1995)1985 | 4 (Part I) |
| Ethchlorvynol | UN1971 (IV) | (S.I. 1995)1985 | 3 |
| Ethinamate | UN1971 (IV) | (S.I. 1995)1985 | 3 |
| N-Ethylamphetamine | UN1971 (IV) | (S.I. 2230)1986 | 4 (Part I) |
| Ethyl loflazepate | UN1971 (IV) | (S.I. 1995)1985 | 4 (Part I) |
| Fencamfamin | UN1971 (IV) | (S.I. 2230)1986 | 4 (Part I) |
| Fenethylline | UN1971 (II) | (S.I. 2230)1986 | 2 |
| Fenproporex | UN1971 (IV) | (S.I. 2230)1986 | 4 (Part I) |
| Fludiazepam | UN1971 (IV) | (S.I. 1995)1985 | 4 (Part I) |
| Flunitrazepam | UN1971 (III) | (S.I. 1995)1985 | 3 |
| Flurazepam | UN1971 (IV) | (S.I. 1995)1985 | 4 (Part I) |
| Halazepam | UN1971 (IV) | (S.I. 1995)1985 | 4 (Part I) |
| Haloxazolam | UN1971 (IV) | (S.I. 1995)1985 | 4 (Part I) |
| Ketazolam | UN1971 (IV) | (S.I. 1995)1985 | 4 (Part I) |
| Loprazolam | UN1971 (IV) | (S.I. 1995)1985 | 4 (Part I) |
| Lorazepam | UN1971 (IV) | (S.I. 1995)1985 | 4 (Part I) |
| Lormetazepam | UN1971 (IV) | (S.I. 1995)1985 | 4 (Part I) |
| Mazindol | UN1971 (IV) | (S.I. 1995)1985 | 3 |
| Medazepam | UN1971 (IV) | (S.I. 1995)1985 | 4 (Part I) |
| Mefenorex | UN1971 (IV) | (S.I. 2230)1986 | 4 (Part I) |
| Mephentermine | Not listed | | 3 |
| Meprobamate | UN1971 (IV) | (S.I. 750) 1998 | 3 |
| Mesocarb | UN1971 (IV) | (S.I. 1995)1985 | 4 (Part I) |
| Methyprylone | UN1971 (IV) | (S.I. 1995)1985 | 3 |
| Midazolam | UN1971 (IV) | (S.I. 2589)1990 | 4 (Part I) |
| Nimetazepam | UN1971 (IV) | (S.I. 1995)1985 | 4 (Part I) |
| Nitrazepam | UN1971 (IV) | (S.I. 1995)1985 | 4 (Part I) |
| Nordazepam | UN1971 (IV) | (S.I. 1995)1985 | 4 (Part I) |
| Oxazepam | UN1971 (IV) | (S.I. 1995)1985 | 4 (Part I) |
| Oxazolam | UN1971 (IV) | (S.I. 1995)1985 | 4 (Part I) |
| Pemoline | UN1971 (IV) | (S.I. 1340)1989 | 4 (Part I) |
| Phendimetrazine | UN1971 (IV) | | 3 |
| Phentermine | UN1971 (IV) | (S.I. 1995)1985 | 3 |
| Pinazepam | UN1971 (IV) | (S.I. 1995)1985 | 4 (Part I) |
| Pipradrol | UN1971 (IV) | | 3 |
| Prazepam | UN1971 (IV) | (S.I. 1995)1985 | 4 (Part I) |
| Pyrovalerone | UN1971 (IV) | (S.I. 2230)1986 | 4 (Part I) |
| Temazepam | UN1971 (IV) | (S.I. 1995)1985 | 3 |
| Tetrazepam | UN1971 (IV) | (S.I. 1995)1985 | 4 (Part I) |
| Triazolam | UN1971 (IV) | (S.I. 1995)1985 | 4 (Part I) |

(b)

[All are unlisted in UN Conventions; all are (S.I. 1300) 1996 and all are in Schedule 4 part II of the Regulations]

| Substance or Product | |
| --- | --- |
| Atamestane | Methenolone |
| Bolandiol | Methyltestosterone |
| Bolasterone | Metribolone |
| Bolazine | Mibolerone |
| Boldenone | Nandrolone |
| Bolenol | Norboletone |
| Bolmantalate | Norclostebol |
| Calusterone | Norethandrolone |
| 4-Chloromethandienone | Ovandrotone |
| Clostebol | Oxabolone |
| Drostanolone | Oxandrolone |
| Enestebol | Oxymesterone |
| Epitiostanol | Oxymetholone |
| Ethyloestrenol | Prasterone |
| Fluoxymesterone | Propetandrol |
| Formebolone | Quinbolone |
| Furazabol | Roxibolone |
| Mebolazine | Silandrone |
| Mepitiostane | Stanolone |
| Mesabolone | Stanozolol |
| Mestanolone | Stenbolone |
| Mesterolone | Testosterone |
| Methandienone | Thiomesterone |
| Methandriol | Trenbolone |

(c) (S.I. 1300) 1996: Any compound (not being Trilostane or a compound for the time being specified in sub-paragraph (b) above) structurally derived from 17-hydroxyandrostan-3-one or from 17-hydroxyestran-3-one by modification in any of the following ways, that is to say:

   (i) by further substitution at position 17 by a methyl or ethyl group;
   (ii) by substitution to any extent at one or more of the positions 1, 2, 4, 6, 7, 9, 11 or 16, but at no other position;
   (iii) by unsaturation in the carbocyclic ring system to any extent, provided that there are no more than two ethylenic bonds in any one carbocyclic ring;

    (iv)  by fusion of ring A with a heterocyclic system.

(d)  Any substance which is an ester or ether (or, where more than one hydroxyl function is available, both an ester and an ether) of a substance specified in sub-paragraph (b) or described in sub-paragraph (c) above;

(e)  [All are unlisted in UN Conventions; all are (S.I. 1300)1996 and all are in Schedule 4 Part II of the Regulations]

---

Chorionic Gonadotrophin (HCG)
Clenbuterol
Non-human chorionic gonadotrophin
Somatotropin
Somatrem
Somatropin

---

2. Any stereoisomeric form of a substance for the time being specified in paragraph 1 of this Part of this Schedule, not being phenylpropanolamine.

3. Any salt of a substance for the time being specified in paragraph 1 or 2 of this Part of this Schedule.

4. Any preparation or other product containing a substance for the time being specified in any of paragraphs 1 to 3 of this Part of this Schedule.

*Chapter 3*

# Generic Controls: Miscellaneous

## 3.1  SALTS

The salts of all controlled drugs are also controlled to the same degree as the parent. A salt is the product of reacting a base with an acid. Like many physiologically active chemicals, controlled drugs are mostly bases, often described as nitrogenous bases or, for plant-derived bases, sometimes termed alkaloids. For various reasons, including stability and ease of handling, the salts (especially hydrochlorides, sulfates and phosphates, less commonly organic anions such as tartrates) are more often seen in both commercial and illicit products than the parent substances. Structures 3.1 and 3.2 show two examples of the formation of salts.

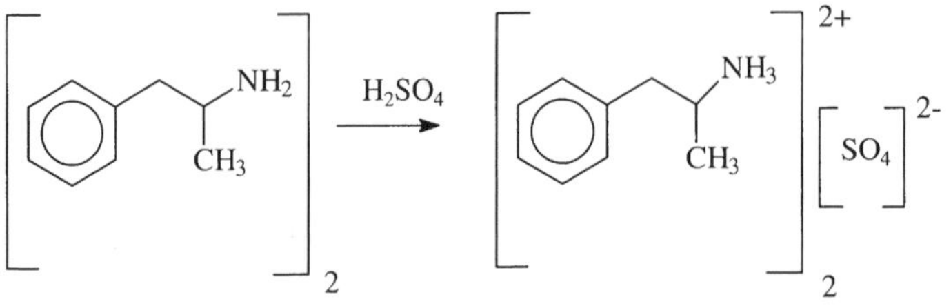

**Structure 3.1**  *The reaction of amphetamine with sulfuric acid to form the sulfate salt*

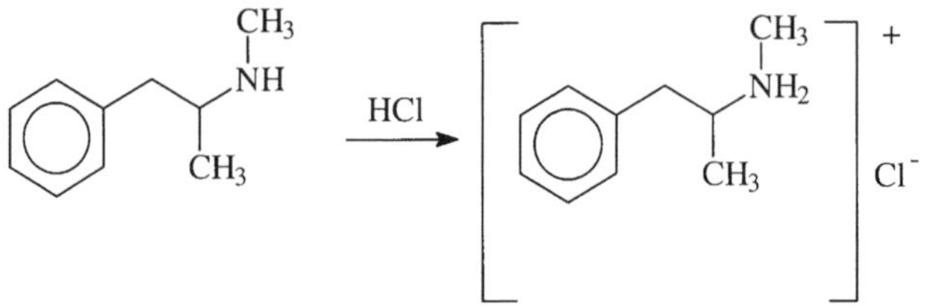

**Structure 3.2**  *The reaction of methylamphetamine with hydrochloric acid to form the hydrochloride salt*

18

Acidic drugs are uncommon; the best examples amongst controlled drugs are the barbiturates. Reaction of a 5,5-disubstituted barbituric acid with sodium hydroxide (a base) produces the sodium salt (see Structure 3.3).

**Structure 3.3** *The reaction of a 5,5-disubstituted barbituric acid with sodium hydroxide to form the disodium salt*

It is not usually necessary for the forensic chemist to identify whether a questioned substance is in its free form (base or acid) or as a particular salt but the need could arise if a case concerned salt–base conversion such as the production of cocaine base (crack) from cocaine hydrochloride (see Appendix 7).

## 3.2   ESTERS AND/OR ETHERS

### 3.2.1   Introduction

The esters and ethers of Class A substances and of the anabolic/androgenic steroids in Class C are subject to the same controls as their unmodified parents, unless that ester or ether is already specified elsewhere in Schedule II. Only structures with a hydroxyl (-OH), sulfhydryl (-SH) or a suitable acid (*e.g.* carboxylic -COOH) group commonly form esters, and only hydroxyl and sulfhydryl groups commonly form ethers. Amongst those basic drugs listed in the Act, which are able to form an ester or an ether, only the hydroxyl function is found.

### 3.2.2   Esters

An example of ester formation is the conversion of morphine to diamorphine (the diacetyl ester of morphine) as shown in Structure 3.4. This process is used, for example, in the illicit production of heroin (crude diamorphine). Diamorphine slowly hydrolyses in damp conditions or rapidly in aqueous alkaline solutions to produce 6-monoacetyl morphine (Structure 3.5). Monoacetylmorphine is still an ester of morphine and therefore remains a Class A controlled drug.

**Structure 3.4**   *The esterification of morphine to diamorphine (Ac₂O is acetic anhydride)*

**Structure 3.5**   *The hydrolysis of diamorphine to form 6-monoacetylmorphine*

Another example of an ester is psilocybin, the naturally-occurring phosphate of psilocin (Structure 3.6). Both psilocin and psilocybin are found in certain fungi of the *Psilocybe* genus (so-called magic mushrooms; see Chapter 8).

The illicit production of the acetyl ester of tetrahydrocannabinol (THC) has been recorded in clandestine laboratories. It is claimed that the resulting THC acetate is a more potent drug.

Esters of the Class C steroids are quite common in commercial formulations. Structures 3.7 and 3.8 show testosterone and its propionate ester. Other common esters of testosterone are the 17$\beta$-cyclopentanepropionate (cypionate) and the 17$\beta$-undecanoate (undecyclate).

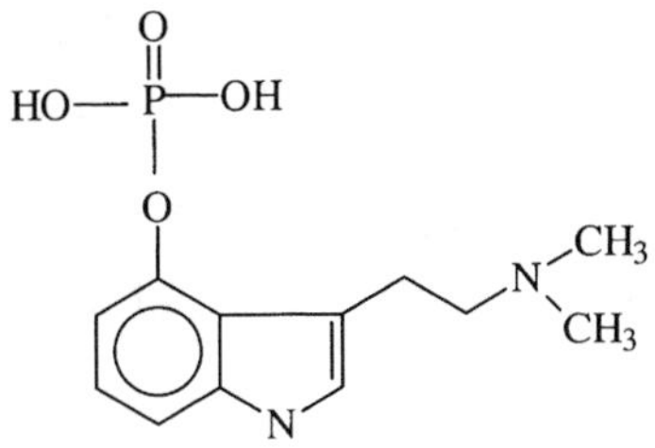

**Structure 3.6**  *Psilocybin, the naturally-occurring phosphate ester of psilocin. The covalent form shown here is in equilibrium with the internal salt*

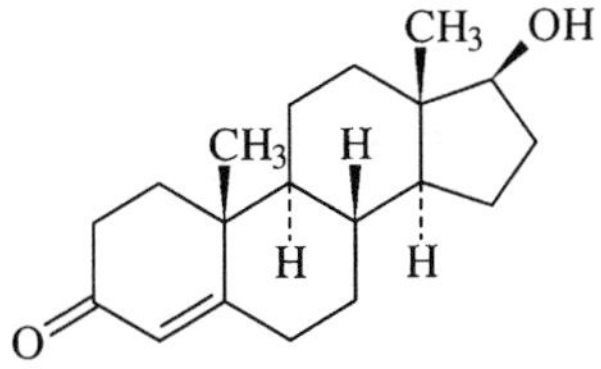

**Structure 3.7**  *Testosterone, an androgenic anabolic steroid*

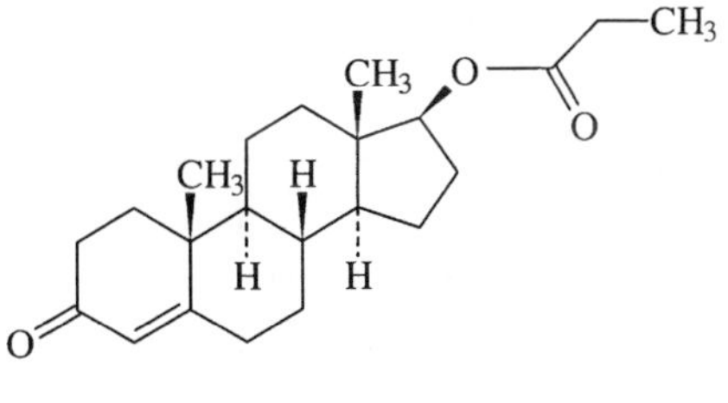

**Structure 3.8**  *The propionate ester of testosterone*

Unlike UN1961, which extended control to esters *and* ethers, the UK legislation specifically refers to esters *or* ethers. In other words, substances which are both esters *and* ethers of a Class A drug or of a Class C steroid are considered to be double derivatives and therefore not controlled unless they are specified elsewhere. An example here is thebacon, which is a named Class A drug. It is an ester and an ether of the Class A drug hydromorphone.

### 3.2.3  Ethers

A number of controlled drugs, principally certain opioids, are ethers. The best example is codeine, the 3-methyl ether of morphine (Structure 3.9). However, neither codeine nor dihydrocodeine (the 3-methyl ether of dihydromorphine) are Class A drugs because they are already listed

under Class B, although a different ether of morphine or of dihydromorphine or of any other suitable Class A drug would be controlled
under Class A.

Various substituted tryptamines, which would qualify as Class A
esters or ethers, are discussed in more detail in Chapter 7.

**Structure 3.9**   *Codeine, the 3-methyl ether of morphine*

## 3.3   STEREOISOMERISM

### 3.3.1   Introduction

Stereoisomers are substances with the same molecular structure, but with
different spatial arrangements of their atoms in the molecule, leading to
different physical and pharmacological properties. Optical stereoisomers
usually contain in their molecule one or more so-called chiral centres.
(An exception to this is found in geometrical stereoisomers which have
no chiral centre. Steric hindrance can produce a rotameric pair, an
example of which is found in the non-controlled drug gossypol).
Different spatial arrangements at one chiral centre give rise to two
molecules that are related as mirror images, and are called enantiomers.
Such molecules are normally optically active (they rotate the plane of
polarised light) and were formerly designated (D) or (+) or as (L) or (−).
They may also form an optically neutral racemate, *i.e.* a mixture of equal
numbers of (+) and (−) molecules, shown as (DL) or (±). These terms
are now obsolete; the present standard designation for steric configuration is the *R* (*rectus*, right) and *S* (*sinister*, left) notation. It is related to
the absolute steric configuration of substituents at chiral centres.
Substances which contain more than one chiral centre give rise to
diastereoisomers.

### 3.3.2   Stereoisomerism in Controlled Drugs

Stereoisomerism is common amongst controlled drugs. In all cases, the
chiral centre involves an asymmetric carbon atom, which is one having

four different substituents. With a few named exceptions, discussed later, all stereoisomers of controlled drugs are also controlled. As with salts, there is no need for the prosecution to name a particular stereoisomeric form. Structure 3.10 shows the two enantiomers of amphetamine.

**Structure 3.10**   *The two enantiomers of amphetamine*

Whereas enantiomeric pairs are common, there are far fewer instances of controlled drugs with two or more chiral centres. A good example of diastereoisomerism occurs in 1-hydroxy-1-phenyl-2-aminopropane. Here, there are two asymmetric carbon atoms giving rise to four diastereoisomers as set out in Structures 3.11(1) to 3.11(4). The 1*S*,2*S* isomer, (+)norpseudoephedrine, is also known as cathine (Class C). However, the 1*S*,2*R* and 1*R*,2*S* isomers, *i.e.* (+) and (−)norephedrine respectively, when present as a racemic mixture, are known as phenylpropanolamine: a decongestant drug. Phenylpropanolamine is excluded from control by paragraph 2 of Part III of Schedule 2 to the Act. Although not directly relevant to the Act, norephedrine was recently added to Table I of UN1988 because of its use as a precursor to amphetamine.

Further exceptions for certain stereoisomers are made in the Act for dextromethorphan and dextrorphan. These are both of clinical value

**Structure 3.11**   *The four diastereoisomers of 1-hydroxy-1-phenyl-2-aminopropane. (1) (+)Norpseudoephedrine (1S,2S); (2) (−)Norpseudoephedrine (1R, 2R); (3) (+)Norephedrine (1S,2R); (4) (−)Norephedrine (1R,2S). The 1S,2S configuration is cathine (Class C); the 1S,2R and 1R,2S racemate is phenylpropanolamine*

although dextromethorphan is occasionally misused. Their enantiomers (*i.e.* levomethorphan and levorphanol respectively) have much greater abuse potential and are both Class A controlled drugs. An extreme case of stereoisomeric complexity amongst controlled drugs occurs with pentazocine, which has three chiral centres and therefore three pairs of diastereoisomers and three racemates.

*Chapter 4*

# Generic Controls: Substa███ ███ ████ ecific

## 4.1  CANNABINOLS

The main psychoactive principal in cannabis is $\Delta^9$-tetrahydrocannabinol
(THC); the concentration in imported herbal cannabis and resin is
typically 5%. Other major components are cannabinol and cannabidiol.
Cannabinol (except where contained in cannabis or cannabis resin) and
certain specified cannabinol derivatives are Class A drugs. As with
cannabis and cannabis resin, it is possible that cannabinols will be moved
to Class C (see Chapter 9).

In Part IV of Schedule 2, 'cannabinol derivatives' are described as
'...the following substances, except where contained in cannabis or
cannabis resin, namely tetrahydro derivatives of cannabinol and 3-alkyl
homologues of cannabinol or of its tetrahydro derivatives'. Structure 4.1
shows cannabinol and Structure 4.2 shows tetrahydrocannabinol. The
IUPAC-preferred ring-numbering system is based on the dibenzopyran
system rather than the earlier monoterpene system. A controlled drug
arises if, in Structures 4.1 or 4.2, the substituent at R is an alkyl group
with six or more carbon atoms (see Chapter 6 for a discussion of
homologues). This definition excludes, for example, the cannabivarins
and the natural carboxylic acid precursors to cannabinols.

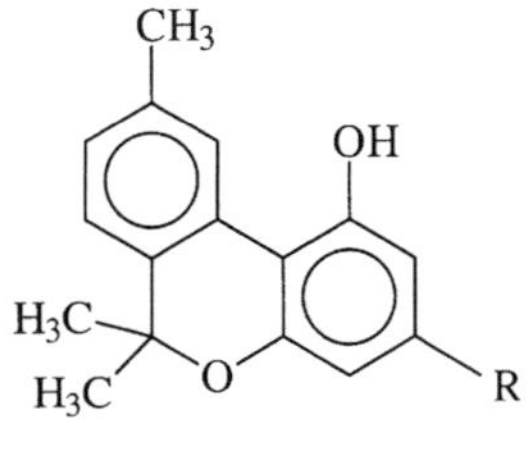

**Structure 4.1**  *Cannabinol (R = pentyl)*

25

**Structure 4.2**    *Tetrahydrocannabinol (R = pentyl). The ring-numbering system is based on the dibenzopyran system*

Controlled cannabinols are covered by Schedule 1 of the Misuse of Drugs Regulations 2001. Research is currently underway to evaluate the clinical potential of a number of cannabinoids extracted from cannabis plants. In the meantime, nabilone, a synthetic analogue of THC, is licensed for hospital use in the UK as a treatment for the nausea arising from cancer chemotherapy, but not for other conditions. Dronabinol, a pharmaceutical product containing the *R,R-trans* stereo-isomer of $\Delta^9$-THC, is not licensed for use in the UK, but may be imported on a 'named patient basis', again for the same indications as nabilone. Dronabinol is marketed as 'Marinol' in the USA. Whereas nabilone is not a controlled drug in the UK, dronabinol, including its stereoisomers, is treated similarly to THC in the Act. But whereas THC is listed in Schedule 1, dronabinol is in Schedule 2 of the Regulations. The possibility exists that a forensic analyst could be required to distinguish the various isomers if a dispute ever arose about the correct scheduling of a questioned sample of $\Delta^9$-THC.

## 4.2   ECGONINE DERIVATIVES

Ecgonine is listed as a Class A drug in paragraph 1(a) of Part I of Schedule 2 to the Act. The full entry reads 'Ecgonine, and any derivative of ecgonine which is convertible to ecgonine or to cocaine'. The relationship between ecgonine and cocaine is shown in Structures 4.3 and 4.4. It will be seen that they differ at both the 2- and 3-positions of the tropane ring.

**Structure 4.3**   *Ecgonine*

**Structure 4.4** *Cocaine*

Bearing in mind the definition of a derivative given in Chapter 6, Structure 4.5 shows benzoylecgonine: a substance which would qualify as a controlled derivative of ecgonine because it can be converted to cocaine by esterification and converted to ecgonine by hydrolysis. Structure 4.6 shows a substance under development as a potential antidepressant drug. Although the synthesis of this compound starts from cocaine, the generally-held view is that it does not qualify as a controlled drug because it cannot be converted to either ecgonine or cocaine in a single stage.

**Structure 4.5** *A controlled derivative of ecgonine (benzoylecgonine)*

**Structure 4.6** *A non-controlled derivative of ecgonine*

## 4.3   LYSERGIDE AND DERIVATIVES OF LYSERGAMIDE

Structure 4.7 shows lysergamide with all nitrogen substituents identified. Lysergide (LSD) is named specifically as a Class A drug; it is the diethylamide of lysergic acid where the ethyl groups are at $R'$

and R″. Generic control is extended to 'other *N*-alkyl derivatives of lysergamide'. A Class A drug therefore arises if R′ and/or R″ and/or R¹ is alkyl.

**Structure 4.7**    *Lysergamide (R′ = R″ = R¹ = H; R⁶ = methyl) showing substitution patterns at the three nitrogen atoms*

It might be questioned whether lysergamide alkylated at R′ and/or R″ and where $R^6$ is an alkyl group other than methyl, would be controlled. This is a 'derivative of a derivative' argument. If $R^6$ is not methyl, then the core structure ought no longer to be regarded as 'lysergamide', and none of its *N*-alkyl derivatives would fall to control. It is believed that the original legislation was not intended to control 1-alkyl lysergamide alkanolamide derivatives such as methysergide, a drug used to treat migraine, for which $R^1$ = methyl, $R^6$ = methyl, R′ = CH(CH$_2$OH)-CH$_2$CH$_3$ and R″ = H.

*Chapter 5*

# Generic Controls: Group-specific

## 5.1 OVERVIEW

The following sections cover the generic classification of phenethyl-amines, tryptamines, fentanyls, pethidines, barbiturates and anabolic steroids. As discussed in Chapter 7, the generic definition of ring-substituted phenethylamines has proved to be remarkably far-sighted. Although far fewer 'designer' tryptamines have been found, the generic controls have again been useful. The controls have also been successful in a reverse sense in that they have not been compromised by or hindered the subsequent development in the pharmaceutical industry of novel compounds for legitimate clinical use. However, the definitions of other groups (fentanyls, pethidines, barbiturates and anabolic steroids) have hardly been tested.

There are no current plans to extend the scope of generic controls. At first sight, it might seem that the benzodiazepines could also be brought within the scope of a structural classification. How-ever, even excluding brotizolam (a thienotriazolodiazepine and not strictly a benzodiazepine) and clotiazepam (a thienodiazepine), the remaining 32 benzodiazepines in the Act still do not form a clear homogeneous group amenable to a readily-comprehensible group definition.

The generic definition of 5,5-disubstituted barbituric acids enables an unknown substance to be uniquely assigned to this group by measuring its UV absorption spectra at pH values corresponding to the formation of a dianion, a monoanion and a neutral species. None of the other group-specific generic definitions provides comparable analytical flex-ibility. Just as with named drugs, the forensic chemist will need to identify unambiguously a generically-controlled substance in all other cases.

## 5.2  PHENETHYLAMINES

In Part I of Schedule 2, paragraph 1(c) defines Class A phenethylamines as: 'any compound (not being methoxyphenamine or a compound for the time being specified in sub-paragraph (a) above) structurally derived from phenethylamine, an *N*-alkylphenethylamine, α-methylphenethylamine, an *N*-alkyl-α-methyl-phenethylamine, α-ethylphenethylamine, or an *N*-alkyl-α-ethylphenethylamine by substitution in the ring to any extent with alkyl, alkoxy, alkylenedioxy or halide substituents, whether or not further substituted in the ring by one or more other univalent substituents.'

Structure 5.1 shows a substituted phenethylamine.

**Structure 5.1**  *Phenethylamine (2-phenylethylamine) showing substitution patterns*

To qualify as a Class A drug, the following criteria must be satisfied:

$R' = $ H or alkyl

$R'' = R^{\alpha 1} = R^{\beta 1} = R^{\beta 2} = $ H

$R^{\alpha 2} = $ H, methyl or ethyl

$R^{x} = $ alkyl, alkoxy, alkylenedioxy or halogen (either singly or in any combination) with or without any other substituents in the ring.

The focus of this rather daunting definition is ring-substitution in amphetamine-like molecules. The reasoning behind this is that the attachment of certain other atoms (especially oxygen, sulfur or halogen) to one or more of the carbon atoms in the aromatic ring of phenethyl-amine leads to major changes in pharmacological properties. Whilst amphetamine and many of its side-chain isomers and other simple derivatives (*e.g.* methylamphetamine, methcathinone, benzphetamine) are all central nervous system stimulants, suitable substitution in the ring can create hallucinogens (*e.g.* mescaline) or empathogenic/entactogenic agents (*e.g.* MDMA) which may or may not retain some stimulant activity. The specific exception of methoxyphenamine was made because this drug, a prescription bronchodilator in the proprietary product

Orthoxine, would have fallen to control under the subsequent definition. However, methoxyphenamine was withdrawn from general use in the UK in 1986 and its continued exception in the legislation is now redundant. The generic definition deliberately excludes from control ring-hydroxy phenethylamines, a group which includes naturally-occurring products such as dopamine, tyramine and adrenaline as well as clinically useful substances such as 4-hydroxyamphetamine. Fenfluramine (*N*-ethyl-α-methyl-3-trifluoromethylphenethylamine), an anorectic drug, is also excluded from control; it has a haloalkyl ring-substitution which is neither explicitly halide nor alkyl (*i.e.* it is a 'derivative of a derivative').

Structures 5.2 to 5.6 show examples of phenethylamines, some of which do and some do not qualify as controlled drugs under the generic definition. Structure 5.2 depicts the well-known example of MDMA (3,4-methylenedioxymethylamphetamine) which qualifies as a Class A controlled drug. The substance in Structure 5.3, one of the 'PIHKAL' drugs, fails the generic test, but is listed by name in paragraph 1(ba) of Part I of Schedule 2 as 4-bromo-β,2,5-trimethoxyphenethylamine (see Chapter 7). Structure 5.4 shows an example of a non-controlled phenethylamine (*N,α*-dimethyl-4-nitrophenethylamine); it is not listed either specifically or generically, but has been described as having pharmacological properties similar to those of analogous phenethylamines. Structure 5.5 shows methoxyphenamine (*o*-methoxymethylamphetamine) which is specifically excluded from control, while Structure 5.6 shows its *p*-isomer (PMMA): a substance that falls within the generic definition. PMMA (*para*methoxymethylamphetamine; methyl-MA) has been seen in drug seizures in Europe and was subjected to risk assessment by the EMCDDA in late 2001 (see Chapter 1). Although their mass spectra, for example, do show some small differences, the forensic analyst would need to ensure that methoxyphenamine and PMMA could be clearly distinguished from each other.

**Structure 5.2** *MDMA (3,4-methylenedioxymethylamphetamine): a Class A substance controlled by the generic definition of a substituted phenethylamine*

**Structure 5.3**   *A phenethylamine derivative from the 'PIHKAL' list (4-bromo-β,2,5-trimethoxyphenethylamine), not covered by the generic definition of a substituted phenethylamine, but listed by name as a Class A drug*

**Structure 5.4**   *A non-controlled phenethylamine (N,α-dimethyl-4-nitrophenethylamine), not covered by the generic definition of a substituted phenethylamine and not listed by name in the Act*

**Structure 5.5**   *Methoxyphenamine, specifically excluded from control*

**Structure 5.6**   *Paramethoxymethylamphetamine (PMMA), the p-isomer of methoxyphenamine and covered by the generic definition*

## 5.3   TRYPTAMINES

In Part I of Schedule 2, paragraph 1(b) defines Class A tryptamines as: 'any compound (not being a compound for the time being specified in sub-paragraph (a) above) structurally derived from tryptamine or from a ring-hydroxy tryptamine by substitution at the nitrogen atom of the side-chain with one or more alkyl substituents but no other substituent'. However, in Part I of Schedule 2, paragraph 3 also provides for any ester or ether to be controlled. Taking both of these requirements together, then to qualify as a Class A drug, the following criteria must be satisfied (see Structure 5.7):

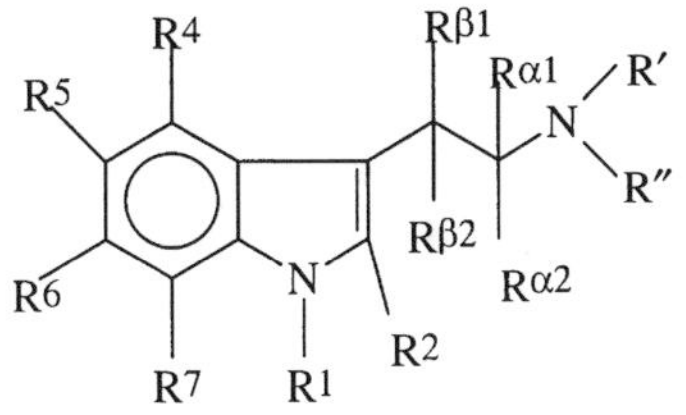

**Structure 5.7** *Tryptamine showing substitution patterns*

$R^4$, $R^5$, $R^6$ and $R^7 = $ H, OH, $OR^x$ or $O(CO)R^x$ where $R^x$ includes alkyl or aryl

$R^1 = R^2 = R^{\alpha 1} = R^{\alpha 2} = R^{\beta 1} = R^{\beta 2} = $ H

$R' = $ H or alkyl

$R'' = $ alkyl

Examples of tryptamines which do and do not satisfy these criteria are discussed in detail in Chapter 7. The generic definition deliberately excludes from control tryptamines where there is no alkyl substitution on the side-chain nitrogen: a group which includes naturally-occurring products such as serotonin and tryptamine itself.

## 5.4   FENTANYLS

In Part I of Schedule 2, paragraph 1(d) defines Class A fentanyls as: 'any compound (not being a compound for the time being specified in sub-paragraph (a) above) structurally derived from fentanyl by modification in any of the following ways, that is to say:

   (i)  by replacement of the phenyl portion of the phenethyl group by any heteromonocycle whether or not further substituted in the heterocycle;

  (ii)  by substitution in the phenethyl group with alkyl, alkenyl, alkoxy, hydroxy, halogeno, haloalkyl, amino or nitro groups;

 (iii)  by substitution in the aniline ring with alkyl, alkoxy, alkylene-dioxy, halogeno or haloalkyl groups;

 (iv)  by substitution at the 4-position of the piperidine ring with any alkoxycarbonyl or alkoxyalkyl or acyloxy group;

  (v)  by replacement of the *N*-propionyl group by another acyl group.'

Structure 5.8 shows fentanyl upon which the above rules operate. There are four named fentanyl derivatives in Class A. Two of these (lofentanil and carfentanil) are not covered by the above definition; they were added to Part I of Schedule 2 in 1988 at the same time as the above generic definition was introduced. The other two (alfentanil and sufentanil) are covered by the above defintion. Remifentanil will shortly be added to the Act as a named Class A drug (see Chapter 9); it does not comply with the above rules.

**Structure 5.8** *Fentanyl*

## 5.5 PETHIDINES

In Part I of Schedule 2, paragraph 1(e) defines Class A pethidines as: 'any compound (not being a compound for the time being specified in sub-paragraph (a) above) structurally derived from pethidine by modification in any of the following ways, that is to say:

  (i) by replacement of the 1-methyl group by an acyl, alkyl whether or not unsaturated, benzyl or phenethyl group, whether or not further substituted;
 (ii) by substitution in the piperidine ring with alkyl or alkenyl groups or with a propano bridge, whether or not further substituted;
(iii) by substitution in the 4-phenyl ring with alkyl, alkoxy, aryloxy, halogeno or haloalkyl groups;
(iv) by replacement of the 4-ethoxycarbonyl by any other alkoxy-carbonyl or any alkoxyalkyl or acyloxy group;
 (v) by formation of an *N*-oxide or of a quaternary base.'

Structure 5.9 shows pethidine upon which the above rules operate. There are a number of Class A drugs closely related to pethidine. Three pethidine intermediates are listed by name in the Act since none complies with the above definition. They are 4-cyano-1-methyl-4-phenylpiper-idine, 1-methyl-4-phenylpiperidine-4-carboxylic acid and 4-phenylpiper-idine-4-carboxylic acid ethyl ester. Four other substances also fail to meet the above rules (difenoxin, diphenoxylate, phenoperidine and hydroxy-pethidine). However, five older named Class A drugs are subsumed by the generic definition (allylprodine, alphameprodine, alphaprodine, proper-idine and trimeperidine).

**Structure 5.9** *Pethidine*

## 5.6 BARBITURATES

In Part II of Schedule 2, paragraph 1(c) defines Class B barbiturates as: 'any 5,5-disubstituted barbituric acid'. This definition only covers the 2-oxo series, so the 2-thio series (*i.e.* thiobarbiturates such as thiopentone) are excluded from control. Structure 5.10 shows a 5,5-disubstituted barbituric acid. In practice, all clinically-useful barbiturates have $R^{5\alpha}$ and $R^{5\beta}$ = alkyl, alkenyl or aryl. The 1,5,5-trisubstituted barbiturate, methylphenobarbitone, does not comply with this rule and is therefore listed specifically as a Class B drug. The twelve barbiturates listed in UN1971 are shown in Table 5.1; this illustrates the economy of the UK generic definition.

**Structure 5.10** *A 5,5-disubstituted barbituric acid*

Most barbiturates are covered by Schedule 3 of the Regulations, but quinalbarbitone (secobarbital), which is also 5,5-disubstituted, is named specifically in Schedule 2 of the Regulations. This is partly because of its higher intrinsic toxicity. Barbitone (5,5-diethylbarbituric acid), a substance often used in buffering solutions, clearly qualifies as a Class B drug. But the Regulations exempt 'a person in charge of a laboratory when acting in his capacity as such' from the restrictions on possession and supply of barbitone (and any other Schedule 3 drug) when in the form of buffering solutions.

**Table 5.1**  *The twelve barbiturates listed in UN1971(see structure 5.10)*

| Name[(i)] | $R^1$ | $R^{5\alpha}$ | $R^{5\beta}$ |
|---|---|---|---|
| Allobarbital | H | Allyl | Allyl |
| Amobarbital | H | Ethyl | Isopentyl |
| Barbital | H | Ethyl | Ethyl |
| Butalbital | H | Allyl | Isobutyl |
| Butobarbital | H | Ethyl | *n*-Butyl |
| Cyclobarbital | H | Ethyl | 1-Cyclohexen-1-yl |
| Methylphenobarbital | Methyl | Ethyl | Phenyl |
| Pentobarbital | H | Ethyl | 1-Methyl-(*n*)-butyl |
| Phenobarbital | H | Ethyl | Phenyl |
| Secbutabarbital | H | Ethyl | *sec*-Butyl |
| Secobarbital | H | Allyl | 1-Methyl-(*n*)-butyl |
| Vinylbital | H | Vinyl | 1-Methyl-(*n*)-butyl |

[(i)]The name used in UN1971 is the Recommended International Non-proprietary Name (rINN).

## 5.7  ANABOLIC STEROIDS

In Part III of Schedule 2, paragraph 1(c) defines Class C anabolic/ androgenic steroids as: 'any compound (not being Trilostane or a compound for the time being specified in sub-paragraph (b) above) structurally derived from 17-hydroxyandrostan-3-one or from 17-hydroxyestran-3-one by modification in any of the following ways, that is to say:

(i)  by further substitution at position 17 by a methyl or ethyl group;
(ii) by substitution to any extent at one or more of the positions 1, 2, 4, 6, 7, 9, 11 or 16, but at no other position;
(iii) by unsaturation in the carbocyclic ring system to any extent, provided that there are no more than two ethylenic bonds in any one carbocyclic ring;
(iv) by fusion of ring A with a heterocyclic system.'

Structures 5.11 and 5.12 show the general structure of the two steroids upon which the above rules operate. Both the 17α and the 17β

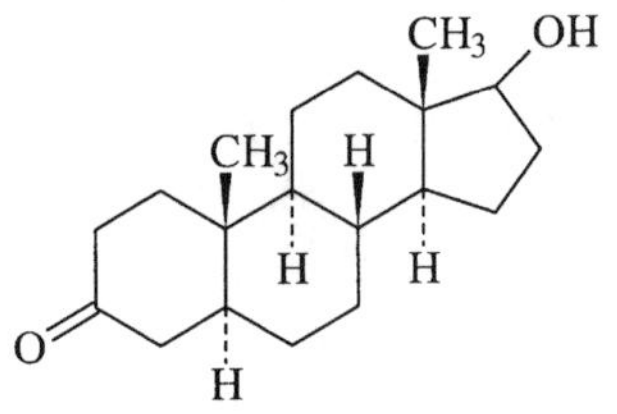

**Structure 5.11**  *17-Hydroxyandrostan-3-one*

**Structure 5.12**   *17-Hydroxyestran-3-one*

**Structure 5.13**   *Trilostane, specifically excluded from control*

configurations are included in the control. Four further anabolic steroids will shortly be added to the Act as named Class C drugs (see Chapter 9). Trilostane (Structure 5.13), which would otherwise be included in the generic definition, is specifically excluded since it has clinical value as an adrenocortical suppressant used in the treatment of breast cancer.

*Chapter 6*

# Nomenclature

## 6.1   BRITISH APPROVED NAMES AND INTERNATIONAL NON-PROPRIETARY NAMES

In common with most chemical substances, a given drug may have a number of synonyms. Wherever possible, the Act uses the British Approved Name (BAN), which is defined by the British Pharmacopoeia Commission. In general, a BAN only exists for drugs that have or have had clinical value. If no BAN exists, then the formal chemical name may be used, *i.e.* a name in agreement with the rules of the International Union of Pure and Applied Chemistry (IUPAC).

There is no intention at present to ensure that the names of drugs in the Act correspond with Recommended International Non-proprietary Names (rINN). However, it should be noted that European Law (Directives 65/65 and 92/27/EEC) requires the use of an rINN for the labelling of medicinal products. The assignment of an rINN to a substance is decided by the WHO. In many cases, the differences between a BAN and an rINN are minor, but in others are substantial. Table 6.1 sets out those controlled drugs where the name used in the Act differs from the rINN. The introduction of International Non-proprietary Names into the Act would not be without its problems. For example, there are many occasions where the rINN refers to a specific stereoisomer, even though the names in the Act must necessarily include all stereoisomers. Furthermore, at least one hybrid name exists in the Act. Thus *N*-hydroxy-tenamphetamine is neither a BAN, an rINN nor an acceptable IUPAC name. Even the core word 'tenamphetamine' itself is an anglicised version of 'tenamfetamine', the Recommended International Non-proprietary Name for one of the 'Ecstasy' drugs commonly called 3,4-methylenedioxyamphetamine (MDA).

**Table 6.1** *Controlled drugs where the name used in the Act differs from the Recommended International Non-proprietary Name (rINN)*

| Name in Act | Class | rINN |
|---|---|---|
| Dimenoxadole | A | Dimenoxadol |
| Methadyl acetate | A | Acetylmethadol |
| Tilidate | A | Tilidine[i] |
| 4-Bromo-2,5-dimethoxy-α-methylphenethylamine | A | Brolamfetamine |
| *N*-Hydroxy-tenamphetamine | A | *N*-Hydroxytenamfetamine[ii] |
| Amphetamine | B | Amfetamine |
| Methylamphetamine | B | Methylamfetamine[iii] |
| Methylphenobarbitone | B | Methylphenobarbital |
| Quinalbarbitone | B | Secobarbital |
| Benzphetamine | C | Benzfetamine |
| Clorazepic acid | C | Dipotassium clorazepate[iv] |
| Diethylpropion | C | Amfepramone[i] |
| Fenethylline | C | Fenetylline |
| Methyprylone | C | Methyprylon |
| *N*-Ethylamphetamine | C | Etilamfetamine |
| Ethyloestrenol | C | Ethylestrenol |
| Methandienone | C | Metandienone |
| Methenolone | C | Metenolone |
| Stanolone | C | Androstanolone |

[i]Amfepramone and Tilidine are proposed International Non-proprietary Names (pINN).
[ii]*N*-Hydroxy-tenamphetamine is not an rINN, but this construction would seem appropriate by analogy to other 'amfetamines'.
[iii]Methylamfetamine is strictly the rINN for the D-isomer only.
[iv]Dipotassium clorazepate is a salt of clorazepic acid and a modified International Non-proprietary Name (INNM). Since salts are already subsumed in the Act, it would be more logical to retain the name as clorazepic acid.

## 6.2 SYNONYMS AND COMMON TERMS

Table 6.2 gives examples of drug names which either do not occur in the Act as such because of generic definitions, or there is a better-known abbreviation, or the trivial name is more widely used. In other cases, US English offers alternative spellings or there are acceptable chemical synonyms. A large number of slang terms for drugs are also in public use although their popularity varies from place to place and in time; a few are shown in Table 6.2. Apart from 'Bromo-STP' and 'STP' shown below, a large number of other controlled phenethylamines (mainly the so-called PIHKAL compounds) are frequently described by acronyms (see Chapter 7). A further example of synonymy occurs in the generic definitions where halide [paragraph 1(c)] and halogeno [paragraph 1(d) of Part I of Schedule 2] should be interpreted as having the same meaning.

**Table 6.2** *Synonyms and other names for certain controlled drugs*

| Common Name | Name in Act | Alternative Name | Slang Terms |
|---|---|---|---|
| Bromo-STP, DOB | 4-Bromo-2,5-dimethoxy-α-methyl-phenethylamine | Brolamfetamine; 4-Bromo-2,5-dimethoxyamphetamine | |
| Cannabis | Cannabis | Marijuana (US), Hemp | Pot, Dope, Blow, Weed |
| Cannabis resin | Cannabis resin | Hashish (US) | Ganga, Charas |
| Heroin[i] | Diamorphine[i] | Diacetylmorphine | 'H', Horse, Skag, Smack |
| LSD | Lysergide | Lysergic acid diethylamide | LSD-25, Acid |
| MDEA | (Generically subsumed) | 3,4-Methylenedioxyethylamphetamine | 'E', Eve, Ecstasy |
| MDMA | (Generically subsumed) | 3,4-Methylenedioxymethylamphetamine | 'E', Adam, Ecstasy, XTC |
| Methcathinone | Methcathinone | Ephedrone | Crank |
| Methylamphetamine | Methylamphetamine | Methamphetamine (US) | Meth, Speed, Ice |
| N-hydroxy MDA | N-Hydroxy-tenamphetamine | N-Hydroxytenamfetamine (US) | |
| Pethidine | Pethidine | Meperidine (US) | |
| STP | 2,5-Dimethoxy-α,4-dimethyl-phenethylamine | 2,5-Dimethoxy-4-methylamphetamine | |

[i]Heroin and diamorphine are not strictly synonymous. The former is a crude preparation in which diamorphine is often the major component.

## 6.3  REDUNDANCY

The introduction, from the late 1970s, of generic definitions based on chemical substitution patterns (Chapter 5) led to a certain amount of duplication. In other words, some controlled drugs continued to be named specifically, but also fell within the scope of generic control. However, this situation was not new; several examples of redundancy can be traced back to the 1972 Protocol, which revised UN1961. Amongst others, this extended controls to include the esters and ethers of substances in Schedule I (see also Chapter 3). The best example of added redundancy is that both morphine and diamorphine continue to be listed by name. However, the explicit retention of diamorphine, the diacetyl ester of morphine, is not strictly required. It is unlikely that the authors of that legislation were unaware of this duplication and it can be argued that no harm is caused. Indeed removal of the word 'diamorphine', while having no legal significance, might be misunderstood and lead to unnecessary debate. Apart from the diamorphine/ morphine pair, a number of other opioids continue to be listed in Schedule I of UN1961 and as Class A controlled drugs in the Act even though they were either ethers or esters of other listed substances. A further level of duplication is caused by the retention in Schedule I of UN1961, and in Part I of Schedule 2 to the Act, of three racemic forms. Thus racemoramide, racemethorphan and racemorphan are redundant because they each contain 50% of an existing controlled drug, namely dextromoramide, levomethorphan and levorphanol respectively. Table 6.3 shows examples of Class A substances where effective double entry occurs in the Act.

## 6.4  DERIVATIVES

### 6.4.1  The Meaning of 'Derivative' in the Act

The concept of a derivative enters the legislation in several places; and the examples of cannabinol, ecgonine and lysergamide derivatives are described in Chapter 4. Pentavalent nitrogen morphine derivatives are also included as Class A drugs, but, like the named morphine *N*-oxide, they are uncommon. Since all organic compounds could be described as derivatives of methane, one of the simplest of all carbon-based compounds, then it follows that they are all derivatives of each other. This *reductio ad absurdum* shows that the word must be given a much tighter meaning when interpreting the Act. It is now usually accepted amongst forensic chemists that compound **A** is a derivative of compound

**Table 6.3** *'Double entry' of Class A controlled drugs in the Act*

| Specific Name (*Paragraph 1(a) of Part I of Schedule 2*) | Generic Control in Part I of Schedule 2 |
|---|---|
| *N,N*-Diethyltryptamine <br> *N,N*-Dimethyltryptamine | Paragraph 1(b) – as *N,N*-dialkyl tryptamines without ring-substitution |
| Bufotenine <br> Psilocin | Paragraph 1(b) – as *N,N*-dialkyl tryptamines with ring-hydroxy substitution |
| 4-Bromo-2,5-dimethoxy-α-methylphenethylamine <br> 2,5-Dimethoxy-α, 4-dimethyl phenethylamine <br> Mescaline | Paragraph 1(c) – as ring-substituted phenethylamines |
| Alfentanil <br> Sufentanil | Paragraph 1(d) – as fentanyl derivatives |
| Allylprodine <br> Alphameprodine <br> Alphaprodine <br> Properidine <br> Trimeperidine | Paragraph 1(e) – as pethidine derivatives |
| Diamorphine | Paragraph 3 – as a diester of morphine |
| Methadyl acetate | Paragraph 3 – as an ester of methadone (enol form) |
| Myrophine | Paragraph 3 – as a diester of morphine |
| Nicomorphine | Paragraph 3 – as a diester of morphine |
| Thebacon | Paragraph 3 – as an ester of hydrocodone (enol form) |
| Hydrocodone | Paragraph 3 – as an ether of hydromorphone |
| Levomethorphan | Paragraph 3 – as an ether of levorphanol |
| Oxycodone | Paragraph 3 – as an ether of oxymorphone |

Note: Thebacon is an ester *and* an ether of hydromorphone and is therefore not covered directly by paragraph 3 of Part I of Schedule 2. In other words, hydrocodone and thebacon are not both redundant, but either could be listed without the other.

**B** only if **B** can be converted (even if only hypothetically) to **A** *in a single chemical reaction.*

### 6.4.2 Dialkyl Derivatives

In Modification Order S.I. 3932 of 2001, two phenethylamine derivatives are included (Compounds #18 and #19 in Table 7.3) where the nitrogen atom is disubstituted with alkyl groups. This was necessary because some doubt resided in whether *N,N-di*substitution on the amine is currently subsumed by the generic definition in paragraph 1(c) of Part I of Schedule 2 (*viz.* '... an *N*-alkylphenethylamine..'). However, another instance of '*N*-alkyl' substitution arises in paragraph 1(a) of Schedule 2 Part I. But here the explicit use of the phrase 'Lysergide and other *N*-alkyl derivatives of lysergamide', by focusing on lysergide (a dialkyl derivative), was therefore intended to mean that '*N*-alkyl' subsumes *N,N*-dialkyl.

### 6.4.3   The Meaning of 'Structurally Derived From'

A different concept of 'derivative' occurs in some of the group-generic definitions discussed in Chapter 5. For example, in ring-substituted phenethylamines, reference is made to 'a compound...... tructurally derived from phenethylamine'. In the judgement in the case of R-*v*-Couzens and Frankel in 1992 (see Appendix 7), it was accepted that to say that compound **A** is *structurally derived* from **B** does not necessarily mean that **B** can be chemically converted to **A** in one or even several reaction stages. What is meant in the example of phenethylamines is that **A** still contains the carbon skeleton of phenethylamine (*i.e.* **B**), but that additional atoms (carbon, oxygen or other as defined) are now attached without implying that such an attachment is chemically possible. In practical terms, it will almost always be the case that **A** and **B** are produced from quite separate precursor chemicals which, in this example, may not in themselves be phenethylamines.

### 6.4.4   Homologues

A particular type of derivative is known as a homologue. This term is used to describe a member of a series of chemical substances where each member differs from the next by a constant structural feature. The simplest example occurs with the straight chain alkanes. Thus methane, ethane, *n*-propane, *n*-butane *etc.* are part of a homologous series where the constant difference between adjacent members is a methylene ($CH_2$) moiety. It is therefore correct to say, for example, that ethane, *n*-propane, *n*-butane *etc.* are the higher homologues of methane and that methane and ethane are the lower homologues of *n*-propane. The term homologue can be found in the legislation in respect of cannabinol and its tetrahydro derivatives. Thus in Part IV of Schedule 2, these controlled derivatives are described as '3-alkyl homologues'. Since the Act does not define 'homologues', there is some contention on whether this phrase means both higher and lower homologues. There is a view that only the higher homologues are included, in which case the cannabivarins with a 3-propyl group are not controlled.

## 6.5 'PHENETHYLAMINES', 'PHENYLETHYLAMINES' AND 'AMPHETAMINES'

The generic definition embedded in paragraph 1(c) of Part I of Schedule 2 to the Act refers to certain substitution patterns in phenethylamine. The term 'phenethylamine' is a IUPAC-recognised contraction for 2-phenylethylamine (also known as $\beta$-phenethylamine); it does not refer to the isomeric 1-phenylethylamine (also known as $\alpha$-phenethylamine). Because of this distinction, derivatives of 1-phenylethylamine (see Chapter 7), even when they otherwise satisfy the substitution pattern set out in paragraph 1(c) of Part I of Schedule 2, are therefore not controlled drugs.

Many phenethylamines of interest are $\alpha$-methyl-substituted and it is common practice to refer to them in a non-IUPAC approved shorthand form as amphetamine (*viz.* $\alpha$-methylphenethylamine) derivatives. Similarly, $N,\alpha$-dimethyl-substituted phenethylamines are often named as methylamphetamine derivatives.

*Chapter 7*

# 'Designer' Drugs

## 7.1  HISTORICAL BACKGROUND

Although a few synthetic ring-substituted phenethylamines (*e.g.* DOB; 4-bromo-2,5-dimethoxy-α-methylphenethylamine) had been subject to limited abuse since at least the 1960s, it was not until the 1980s that the phenomenon of so-called designer drugs was fully recognised. Two principal series were originally seen: those based on fentanyl and those structurally derived from pethidine (strictly α-prodine, the reverse ester of 3-methylpethidine). In both series, abuse was largely confined to the US, although 3-methyl- and 4-fluorofentanyl have been reported in Europe in recent years. As narcotic analgesics, these substances offered similar effects to heroin, but in much smaller doses. The 4-substituted fentanyls are typically 4000 times more potent analgesics than morphine. Not surprisingly, the low doses needed led to many accidental and fatal overdoses. The α-prodine series also caused a notorious public health issue when it was found that a synthetic by-product of clandestine synthesis, known as MPTP, produced a rapid chemically-induced Parkinson's disease. Many pethidine and fentanyl derivatives were brought within the scope of the Act by generic controls in 1986.

Starting in the late 1980s, a third and much larger series of illicit designer drugs began to appear, all of which were based on the phenethylamine nucleus. Most have been produced in Europe. Apart from amphetamine, by far the most common of these phenethylamines was MDMA (3,4-methylenedioxymethylamphetamine). Their attraction to youth culture is that they offer a mixture of stimulant and so-called empathogenic/entactogenic properties, and they are seen by users as safe drugs. In the UK, abuse of ring-substituted phenethylamines had been well anticipated by the generic controls of 1977.

The original definition of designer drugs described them as '*Analogues, or chemical cousins, of controlled substances that are designed to produce effects similar to the controlled substances they mimic*'. The clear assumption is that a designer drug is not in itself a controlled substance. This is a rather restrictive definition; the chemical-generic controls in the UK legislation capture many substances that may nevertheless still be thought of as designer drugs. It is better to take a wider view, which does not make any assumptions about legal status. The diversion of precursors towards production of amphetamine, MDMA and related drugs has been partly curtailed by UN1988 and its domestic equivalents. Clandestine synthesis of novel phenethylamine designer drugs is now driven more by the availability of alternative precursor chemicals. This is facilitated by the way in which the phenethylamine nucleus lends itself to molecular manipulation; a range of different precursors can be used to make a single drug and a single precursor can make several different drugs. This clandestine activity has been further supported by a proliferation of 'underground' guides to drug production in the past ten years, many of which now appear on the Internet.

Two particular sub-sets of designer drugs, namely phenethylamines (mostly the 'PIHKAL' group) and tryptamines (the 'TIHKAL' group) are described below.

## 7.2. 'NEW' TRYPTAMINES

### 7.2.1 Overview

Tryptamine, 2-(1*H*-indol-3-yl)ethylamine, is a naturally-occurring metabolite of the amino acid tryptophan, which, in turn, is a constituent of many proteins. Although tryptamine has few significant pharmacological properties, it forms the parent nucleus of a number of hallucinogenic drugs. Some of these are simple derivatives of tryptamine, whereas others are polycyclic structurally-related substances such as $\beta$-carbolines and lysergamides. Many hallucinogenic tryptamines occur naturally in plants, fungi and, occasionally, animals, but others are entirely synthetic or semi-synthetic substances.

A number of tryptamines are controlled by the Act as Class A drugs. Five are listed specifically in paragraph 1(a) of Part I of Schedule 2, namely bufotenine, etryptamine, psilocin, *N*,*N*-diethyltryptamine (DET) and *N*,*N*-dimethyltryptamine (DMT). Two structurally-related substances containing the 2-(indol-3-yl)ethylamine fragment (corresponding to tryptamine) are also explicitly listed, namely lysergamide and lysergide. Other tryptamines are subsumed by the generic definition in paragraph 1(b) of Part I of Schedule 2.

### 7.2.2 'TIHKAL'

A 1997 book following a similar format to the authors' previous publication (PIHKAL) is known by the acronym 'TIHKAL' (see Bibliography); it provides detailed synthetic monographs for 55 tryptamines together with notes on dosages, routes of administration, effects and properties of related compounds. Of these 55 substances, nine are complex molecules containing the structure of tryptamine (*i.e.* four lysergamides, four $\beta$-carbolines and ibogaine). The remaining monographs include tryptamine itself and 45 derivatives (Table 7.1). The monograph for lysergide (LSD; Substance #26 in Table 7.2) includes brief details of 43 other LSD derivatives, but these, the four substituted $\beta$-carbolines and ibogaine (a pentacyclic indole alkaloid) are not considered further. The four lysergamide derivatives are shown in Table 7.2. Lysergamide itself is not listed in TIHKAL, but is included in Table 7.2 for reference. The requirements for a substituted tryptamine (see Structure 7.1) to be captured by the existing generic legislation were described in Chapter 2. Structure 7.2 shows a substituted lysergamide (see also Chapter 4).

**Structure 7.1**   *Tryptamine showing substitution patterns*

There is yet little evidence that youth culture and the dance drug scene are likely to move away in the near future from the use of stimulants and empathogens typified by the phenethylamines towards hallucinogens of the tryptamine family. As well as not being stimulants, a more significant limitation is that many tryptamines are inactive when ingested. In order to produce an effect they must be either smoked, injected or mixed with a monoamine oxidase inhibitor (MAOI). A good example of the latter situation is the hallucinogenic drink known as Ayahuasca or Caapi to certain indigenous people of South America. This is a concoction of plant extracts containing DMT (the hallucinogen) and harmine

**Table 7.1** *Derivatives of tryptamine (the position of substituents is depicted in Structure 7.1)*

| TIHKAL[i] | $R^l$ | $R^{l'}$ | $R^{\alpha 1}$ | $R^{\alpha 2}$ | $R^{\beta 1}$ | $R^{\beta 2}$ | $R^2$ | $R^1$ | $R^{4\ to\ 7}$ | Controlled? |
|---|---|---|---|---|---|---|---|---|---|---|
| 2 | n-But | n-But | H | H | H | H | H | H | H | Yes |
| 3 | Et | Et | H | H | H | H | H | H | H | Yes {DET} |
| 4 | i-Pro | i-Pro | H | H | H | H | H | H | H | Yes |
| 5 | H | H | Me | H | H | H | H | H | 5-MeO | No |
| 6 | Me | Me | H | H | H | H | H | H | H | Yes {DMT} |
| 7 | H | H | Me | H | H | H | Me | H | H | No |
| 8 | Me | Me | Me | H | H | H | H | H | H | No |
| 9 | Pro | Pro | H | H | H | H | H | H | H | Yes |
| 10 | Et | i-Pro | H | H | H | H | H | H | H | Yes |
| 11 | H | H | Et | H | H | H | H | H | H | Yes {Etryptamine} |
| 15 | n-But | n-But | H | H | H | H | H | H | 4-HO | Yes |
| 16 | Et | Et | H | H | H | H | H | H | 4-HO | Yes |
| 17 | i-Pro | i-Pro | H | H | H | H | H | H | 4-HO | Yes |
| 18 | Me | Me | H | H | H | H | H | H | 4-HO | Yes {Psilocin} |
| 19 | Me | Me | H | H | H | H | H | H | 5-HO | Yes {Bufotenine} |
| 20 | Pro | Pro | H | H | H | H | H | H | 4-HO | Yes |
| 21 | Me | Et | H | H | H | H | H | H | 4-HO | Yes |
| 22 | Me | i-Pro | H | H | H | H | H | H | 4-HO | Yes |
| 23 | Me | Pro | H | H | H | H | H | H | 4-HO | Yes |
| 24 | {cyclobutyl} | {cyclobutyl} | H | H | H | H | H | H | 4-HO | No |
| 27 | Me | n-But | H | H | H | H | H | H | 4-HO | Yes |
| 28 | i-Pro | i-Pro | H | H | H | H | H | H | 4,5-MDO[ii] | No |
| 29 | i-Pro | i-Pro | H | H | H | H | H | H | 5,6-MDO | No |
| 30 | Me | Me | H | H | H | H | H | H | 4,5-MDO | No |
| 31 | Me | Me | H | H | H | H | H | H | 5,6-MDO | No |
| 32 | Me | i-Pro | H | H | H | H | H | H | 5,6-MDO | No |
| 33 | Et | Et | H | H | H | H | Me | H | H | No |
| 34 | Me | Me | H | H | H | H | Me | H | H | No |
| 35 | Acetyl | H | H | H | H | H | H | H | 5-MeO | No |

| No. | R1 | R2 | R3 | R4 | R5 | R6 | R7 | Ar | TIHKAL |
|---|---|---|---|---|---|---|---|---|---|
| 36 | Et | Et | H | H | H | H | H | 5-MeO | Yes |
| 37 | i-Pro | i-Pro | H | H | H | H | H | 5-MeO | Yes |
| 38 | Me | Me | H | H | H | H | H | 5-MeO | Yes |
| 39 | Me | i-Pro | H | H | H | H | H | 4-MeO | Yes |
| 40 | Me | i-Pro | H | H | H | H | H | 5-MeO | Yes |
| 41 | Me | i-Pro | H | H | H | H | H | 5,6-di-MeO | Yes |
| 42 | Me | H | H | H | H | H | H | 5-MeO | Yes |
| 43 | {cyclobutyl} | {cyclobutyl} | H | H | H | H | H | 5-MeO | No |
| 45 | Me | Me | H | H | Me | H | H | 5-MeO | No |
| 46 | Me | Me | H | H | Me | H | H | 5-MeS | No |
| 47 | Me | i-Pro | H | H | H | H | H | H | Yes |
| 48 | H | H | Me | H | H | H | H | H | No |
| 49 | H | Et | H | H | H | H | H | H | Yes |
| 50 | H | Me | H | H | H | H | H | H | Yes |
| 52 | {cyclobutyl} | {cyclobutyl} | H | H | H | H | H | H | No |
| 53 | H | H | H | H | H | H | H | H | No |
| 55 | H | Me | Me | H | H | H | H | 5-MeO | No |

[i]TIHKAL refers to monograph numbers in that publication. [ii]MDO = Methylenedioxy.

**Structure 7.2**  *Lysergamide showing substitution patterns at the three nitrogen atoms*

**Table 7.2**  *Derivatives of lysergamide (the position of substituents is depicted in Structure 7.2)*

| TIHKAL[i] | R' | R'' | R$^1$ | R$^6$ | Controlled? |
|---|---|---|---|---|---|
| 1 | Et | Et | H | Allyl | No |
| 12 | Et | Et | H | Et | No |
| 26 | Et | Et | H | Me | Yes (Lysergide) |
| 51 | Et | Et | H | n-Pro | No |
| Lysergamide | H | H | H | Me | Yes |

[i]TIHKAL refers to monograph numbers in that publication.

(the MAOI). Some tryptamines listed in TIHKAL appear to have no pharmacological action by any route.

### 7.2.3  The Control Status of TIHKAL Substances

Tables 7.1 and 7.2 show the status of the tryptamine and lysergamide derivatives under the Act. Of the 55 substances featured in the TIHKAL monographs, then excluding tryptamine itself, the four β-carbolines and ibogaine (none of which is controlled), 29 are either listed specifically or defined generically as Class A drugs. In coming to this conclusion, it should be noted that a cyclobutyl substituent which leads to ring closure at the side-chain nitrogen atom (Compounds #24, #43 and #52 in Table 7.1) is not considered to be a *N,N*-dialkyl derivative; these three compounds are therefore not controlled. On the other hand, a ring-dihydroxy group leading to a ring-diether (Substance #41 in Table 7.1) is regarded as satisfying the requirements. If in the generic definition of a Class A tryptamine, the phrase '*ring-hydroxy tryptamine*' were to be replaced with '*ring-hydroxy or ring-alkylenedioxy tryptamine*', then a further five drugs (Substances #28 to #32) would be controlled. Because of the structural heterogeneity of the remaining compounds, specific nomenclature would be required. Three lysergamide derivatives (in which R$^6$ has been changed) are shown as non-controlled in Table 7.2;

as discussed in Chapter 4, they must be regarded as second-order derivatives of lysergamide.

Few of the tryptamines listed in TIHKAL have any actual or potential value to the pharmaceutical industry, but it should be noted that melatonin (Substance #35 in Table 7.1), although unlicensed in the UK, enjoys some status as a fringe medicine for the treatment of jet-lag and other sleeping disorders. Melatonin would not become a controlled drug by virtue of the extension of the generic definition shown above.

## 7.3 'NEW' RING-SUBSTITUTED PHENETHYLAMINES

### 7.3.1  Introduction

The generic controls on phenethylamines (Chapter 5), which were introduced in 1977, were remarkably far-sighted and comprehensive. Not only did they successfully anticipate the major 'Ecstasy' drugs such as MDMA and its congeners, but the generic definition subsumed nearly all of the many ring-substituted amphetamine-type compounds which would appear in UK drug seizures over the next 24 years. The publication of a book in 1991 ('PIHKAL' – see Bibliography) and its subsequent appearance on the world-wide web alerted clandestine chemists to the possibilities of creating further designer drugs based on the phenethylamine nucleus. That book provides synthetic methods for over 170 substances with notes on their effects and doses. Of these, 34 were not covered by the generic definition of 1977. Since many of the 'PIHKAL' drugs were broadly similar to the well-known 'Ecstasy' compounds and related hallucinogens, HM Government recognised that they presented a potential social problem. Although consideration had been given to extending the generic definition of phenethylamines, two arguments worked against this approach. Firstly, the 34 substances formed a heterogeneous set of chemical structures, the inclusion of which would have needed an elaborate definition. This then risked including substances of current or future interest to the pharmaceutical industry. Secondly, the generic control of 1977 has worked remarkably well with no forensic difficulty, but its enlargement could pose a danger not just of incomprehension, but of creating loopholes by virtue of its complexity. The 34 substances are now listed by name as Class A controlled drugs (see Chapter 2).

### 7.3.2  Methylthioamphetamine

In 1997, a new designer drug appeared in Europe: 4-methylthioamphet-amine (4-MTA). This had not been described in 'PIHKAL', but had been reported in the pharmacological literature as a possible antidepressant drug. A number of seizures of 4-MTA were made by police and customs across Europe in the following two years. At least five fatal poisonings were recorded in the UK alone where 4-MTA had been the direct or indirect cause of death. This substance was one of several so-called 'New Synthetic Drugs' that were notified under the European Union Joint Action (see Chapter 1) and subjected to risk assessment. As a consequence of this evaluation, the European Council of Ministers decided in 1999 that 4-MTA should be controlled in all Member States of the Union. In early 2001, the UN CND meeting in Vienna agreed that 4-MTA should be brought within the scope of UN1971 as a Schedule I substance. In the UK, 4-MTA [shown as α-methyl-4-(methylthio)-phenethylamine] has been listed as a Class A controlled drug. Like the 34 'PIHKAL' drugs, it is not covered by the 1977 generic definition and is therefore named specifically in paragraph 1(ba) of Part I of Schedule 2 of the Act.

### 7.3.3   Structural Classification

Table 7.3 lists the 34 'PIHKAL' compounds as well as 4-MTA firstly as they appear in the Act under their IUPAC names, then as their less formal names and acronyms. Where appropriate, the IUPAC name has been based on the more common root '3,4-methylenedioxyphenyl-Z', where Z is some substituent or other part of the molecule. However, the rules strictly prefer the use of the alternative root '1-(1,3-benzodioxol-5-yl)-Z'. The 35 compounds fall into five structural groups (Table 7.4). Groups 1–3 are sub-divided into those (a) where, according to PIHKAL, positive psychoactive effects may be expected, and those (b),where either no effect was detected, the effect was unpleasant or the dose was unacceptably high. The molecular structures of the 35 compounds are set out in Structures 7.3(1) to 7.3(35). For comparison, the substitution pattern required by the generic rules of 1977 is shown in Structure 5.1 (Chapter 5).

## 7.4   SIDE-CHAIN DERIVATIVES OF PHENETHYLAMINE

As noted above, the phenethylamine nucleus has been a particularly fruitful source of new synthetic substances. Apart from the ring-substituted members (*i.e.* MDMA *etc.*), variations in the side-chain have given rise to several minor series, examples of which have occurred in drug seizures.

### 7.4.1   *N*-substituted Phenethylamines

The *N*-substituted phenethylamines make up a rather mixed group. Some are well-known as established drugs of abuse while others have

**Table 7.3** *The 34 PIHKAL substances and 4-MTA showing synonyms and acronyms*

| No.[i] | IUPAC Name in Paragraph 1(ba) of Part I of Schedule 2 | Synonym[ii] | Acronym | Ref. | Page |
|---|---|---|---|---|---|
| 25 | Allyl(α-methyl-3,4-methylenedioxyphenethyl)amine | 3,4-MD-*N*-Allylamphetamine | MDAL | #101 | 719 |
| 16 | 2-Amino-1-(2,5-dimethoxy-4-methylphenyl)ethanol | 2,5-Dimethoxy-β-hydroxy-4-methylPEA | BOHD | #16 | 498 |
| 17 | 2-Amino-1-(3,4-dimethoxyphenyl)-ethanol | 3,4-Dimethoxy-β-hydroxyPEA | DME | #57 | 609 |
| 20 | Benzyl(α-methyl-3,4-methylene-dioxyphenethyl)amine | 3,4-MD-*N*-Benzylamphetamine | MDBZ | #103 | 721 |
| 02 | 4-Bromo-β,2,5-trimethoxy-phenethylamine | 4-Bromo-2,5,β-trimethoxyPEA | BOB | #13 | 490 |
| 10 | *N*-(4-*sec*-Butylthio-2,5-dimethoxy-phenethyl)-hydroxylamine | 2,5-Dimethoxy-4-(*s*)-butylthio-*N*-hydroxyPEA | HOT-17 | #89 | 685 |
| 26 | Cyclopropylmethyl(α-methyl-3,4-methylene-dioxyphenethyl)amine | 3,4-MD-*N*-Cyclopropylmethyl-amphetamine | MDCPM | #104 | 724 |
| 14 | 2-(4,7-Dimethoxy-2,3-dihydro-1*H*-indan-5-yl)-ethylamine | 2,5-Dimethoxy-3,4-(trimethyl-ene)PEA | 2C-G-3 | #28 | 526 |
| 15 | 2-(4,7-Dimethoxy-2,3-dihydro-1*H*-indan-5-yl)-1-methylethylamine | 2,5-Dimethoxy-3,4-(trimethyl-ene)amphetamine | G-3 | #82 | 674 |
| 05 | 2-(2,5-Dimethoxy-4-methyl-phenyl)cyclopropylamine | 2-(2,5-Dimethoxy-4-methylphenyl)-cyclopropylamine | DMCPA | #56 | 607 |
| 27 | 2-(1,4-Dimethoxy-2-naphthyl)-ethylamine | 1,4-Dimethoxynaphthyl-2-ethylamine | 2C-G-N | #31 | 535 |
| 28 | 2-(1,4-Dimethoxy-2-naphthyl)-1-methylethylamine | 1,4-Dimethoxynaphthyl-2-isopropylamine | G-N | #86 | 681 |
| 09 | *N*-(2,5-Dimethoxy-4-propylthio-phenethyl)-hydroxylamine | 2,5-Dimethoxy-*N*-hydroxy-4-(*n*)-propylthioPEA | HOT-7 | #88 | 683 |

*(continued)*

**Table 7.3** Continued

| No.[i] | IUPAC Name in Paragraph 1(ba) of Part 1 of Schedule 2 | Synonym[ii] | Acronym | Ref. | Page |
|---|---|---|---|---|---|
| 29 | 2-(1,4-Dimethoxy-5,6,7,8-tetrahydro-2-naphthyl)-ethylamine | 2,5-Dimethoxy-3,4-(tetramethylene)-PEA | 2C-G-4 | #29 | 529 |
| 30 | 2-(1,4-Dimethoxy-5,6,7,8-tetrahydro-2-naphthyl)-1-methylethylamine | 2,5-Dimethoxy-3,4-(tetramethylene)-amphetamine | G-4 | #83 | 676 |
| 07 | α,α-Dimethyl-3,4-methylene-dioxyphenethylamine | 3,4-MDPhentermine | MDPH | #116 | 748 |
| 06 | α,α-Dimethyl-3,4-methylene-dioxy-phenethyl(methyl)amine | 3,4-MDMephentermine | MDMP | #113 | 743 |
| 19 | Dimethyl(α-methyl-3,4-methylene-dioxy-phenethyl)amine | 3,4-MD-N,N-Dimethylamphetamine | MDDM | #105 | 725 |
| 08 | N-(4-Ethylthio-2,5-dimethoxy-phenethyl)-hydroxylamine | 2,5-Dimethoxy-4-ethylthio-N-hydroxyPEA | HOT-2 | #87 | 682 |
| 18 | 4-Iodo-2,5-dimethoxy-α-methyl-phenethyl-(dimethyl)amine | 2,5-Dimethoxy-N,N-dimethyl-4-iodoamphetamine | IDNNA | #90 | 687 |
| 12 | 2-(1,4-Methano-5,8-dimethoxy-1,2,3,4-tetrahydro-6-naphthyl)ethylamine | 3,6-Dimethoxy-4-(2-aminoethyl)-benzonorbornane | 2C-G-5 | #30 | 532 |
| 13 | 2-(1,4-Methano-5,8-dimethoxy-1,2,3,4-tetra-hydro-6-naphthyl)-1-methylethylamine | 3,6-Dimethoxy-4-(2-aminopropyl)-benzonorbornane | G-5 | #84 | 676 |
| 32 | 2-(5-Methoxy-2,2-dimethyl-2,3-dihydro-benzo[b]furan-6-yl)-1-methylethylamine | 6-(2-Aminopropyl)-2,2-dimethyl-5-methoxy-2,3-dihydrobenzofuran | F-22 | #80 | 667 |
| 24 | 2-Methoxyethyl(α-methyl-3,4-methylene-dioxy-phenethyl)amine | 3,4-MD-N-(2-Methoxyethyl)-amphetamine | MDMEOET | #112 | 742 |
| 31 | 2-(5-Methoxy-2-methyl-2,3-dihydro-benzo[b]furan-6-yl)-1-methylethylamine | 6-(2-Aminopropyl)-5-methoxy-2-methyl-2,3-dihydrobenzofuran | F-2 | #79 | 664 |
| 04 | β-Methoxy-3,4-methylene-dioxyphenethylamine | β-Methoxy-3,4-MDPEA | BOH | #15 | 496 |

(continued)

**Table 7.3** *Continued*

| No.[(i)] | IUPAC Name in Paragraph 1(ba) of Part I of Schedule 2 | Synonym[(ii)] | Acronym | Ref. | Page |
|---|---|---|---|---|---|
| 34 | 1-(3,4-Methylenedioxybenzyl)-butyl(ethyl)amine | 2-Ethylamino-1-(3,4-MDphenyl)-pentane | ETHYL-K | #78 | 663 |
| 33 | 1-(3,4-Methylenedioxybenzyl)-butyl(methyl)amine | 2-Methylamino-1-(3,4-MDphenyl)-pentane | METHYL-K | #129 | 781 |
| 22 | 2-(α-Methyl-3,4-methylenedioxy-phenethylamino)-ethanol | 3,4-MD-N-(2-Hydroxyethyl)-amphetamine | MDHOET | #107 | 731 |
| 21 | α-Methyl-3,4-methylenedioxy-phenethyl(prop-2-ynyl)amine | 3,4-MD-N-propargylamphetamine | MDPL | #117 | 752 |
| 11 | N-Methyl-N-(α-methyl-3,4-methylene-dioxy-phenethyl)hydroxylamine | N-hydroxy-N-methyl-3,4-MDA | FLEA | #81 | 671 |
| 23 | O-Methyl-N-(α-methyl-3,4-methylene-dioxy-phenethyl)hydroxylamine | 3,4-MD-N-methoxyamphetamine | MDMEO | #111 | 741 |
| 35 | α-Methyl-4-(methylthio)phenethylamine | 4-methylthioamphetamine | 4-MTA | – | – |
| 03 | β,3,4,5-Tetramethoxyphenethylamine | 3,4,5,β-tetramethoxyPEA | BOM | #17 | 500 |
| 01 | β,2,5-Trimethoxy-4-methylphenethylamine | 4-methyl-2,5,β-trimethoxyPEA | BOD | #14 | 492 |

[(i)]The index number in the first column refers to the parenthetical number in Structures 7.3(1) to 7.3(35). Acronym (except 4-MTA), Ref. and Page refer respectively to the code name, monograph number and page in PIHKAL.
[(ii)]The following abbreviations are used in the synonyms: PEA = phenethylamine; MD = methylenedioxy; MDPEA = methylenedioxyphenethylamine; MDA = methylenedioxyamphetamine.

**Table 7.4** *Classification of the 35 substances added to the Act as Class A drugs by the Modification Order of 2001 (S.I. 3932)*

| Structural Group[(i)] | Substitution Pattern | Substances (See Structures 7.3(1) to 7.3(35)) |
|---|---|---|
| 1a | β-Substitution, α,β-disubstitution or α,α-disubstitution | 1–7 |
| 1b | | 16 and 17 |
| 2a | N-Hydroxy, N-alkenyl, N-aryl, N-hydroxyalkyl, N-cyclopropylmethyl, N-alkoxyalkyl or N,N-disubstitution | 8–11 |
| 2b | | 18–26 |
| 3a | Annulated phenethylamines | 12–15 |
| 3b | | 27–32 |
| 4 | α-Substitution beyond ethyl | 33 and 34 |
| 5 | Ring-substitution other than by alkyl, alkoxy, alkylenedioxy or halide | 35 |

[(i)]Groups 1–3 are divided into those (a) where, according to PIHKAL, positive psychoactive effects may be expected, and those (b), where either no effect was detected, the effect was unpleasant or the dose was unacceptably high.

**Structure 7.3** 7.3(1) *to* 7.3(35) *The 34 'PIHKAL' substances and 4-MTA listed in paragraph 1(ba) of Part I of Schedule 2 to the Act (see also Table 7.3).* (1) *β,2,5-Trimethoxy-4-methylphenethylamine;* (2) *4-Bromo-β,2,5-trimethoxyphenethylamine*

(3) *β,3,4,5-Tetramethoxyphenethylamine;* (4) *β-Methoxy-3,4-methylenedioxyphenethylamine*

(5) *2-(2,5-Dimethoxy-4-methylphenyl)cyclopropylamine;* (6) *α,α-Dimethyl-3,4-methylenedioxyphenethyl(methyl)amine*

(7) *α,α-Dimethyl-3,4-methylenedioxyphenethylamine;* (8) N-(*4-Ethylthio-2,5-dimethoxyphenethyl)hydroxylamine*

(9) N-(*2,5-Dimethoxy-4-propylthiophenethyl)hydroxylamine;* (10) N-(*4-sec-Butylthio-2,5-dimethoxyphenethyl)hydroxylamine*

(11) N-*Methyl*-N-(*α-methyl-3,4-methylenedioxyphenethyl)-hydroxylamine;* (12) *2-(1,4-Methano-5,8-dimethoxy-1,2,3,4-tetrahydro-6-naphthyl)ethylamine*

(13) *2-(1,4-Methano-5,8-dimethoxy-1,2,3,4-tetrahydro-6-naphthyl)-1-methylethylamine;* (14) *2-(4,7-Dimethoxy-2,3-dihydro-1H-indan-5-yl)ethylamine*

(15) *2-(4,7-Dimethoxy-2,3-dihydro-1H-indan-5-yl)-1-methylethylamine;* (16) *2-Amino-1-(2,5-dimethoxy-4-methylphenyl)ethanol*

(17) *2-Amino-1-(3,4-dimethoxyphenyl)ethanol;* (18) *4-Iodo-2,5-dimethoxy-α-methylphenethyl(dimethyl)amine*

(19) *Dimethyl(α-methyl-3,4-methylenedioxyphenethyl)amine;* (20) *Benzyl(α-methyl-3,4-methylenedioxyphenethyl)amine*

(21) *α-Methyl-3,4-methylenedioxyphenethyl(prop-2-ynyl)amine;* (22) *2-(α-Methyl-3,4-methylenedioxyphenethylamino)ethanol*

(23) O-*Methyl*-N-*(α-methyl-3,4-methylenedioxyphenethyl)-hydroxylamine;* (24) *2-Methoxyethyl(α-methyl-3,4-methylenedioxyphenethyl)amine*

(25) *Allyl(α-methyl-3,4-methylenedioxyphenethyl)amine;* (26) *Cyclopropylmethyl(α-methyl-3,4-methylenedioxyphenethyl)amine*

(27) *2-(1,4-Dimethoxy-2-naphthyl)ethylamine;* (28) *2-(1,4-Dimethoxy-2-naphthyl)-1-methylethylamine*

(29) *2-(1,4-Dimethoxy-5,6,7,8-tetrahydro-2-naphthyl)ethylamine;* (30) *2-(1,4-Dimethoxy-5,6,7,8-tetrahydro-2-naphthyl)-1-methylethylamine*

(31) *2-(5-Methoxy-2-methyl-2,3-dihydrobenzo[b]furan-6-yl)-1-methylethylamine;* (32) *2-(5-Methoxy-2,2-dimethyl-2,3-dihydrobenzo[b]furan-6-yl)-1-methylethylamine*

(33) *1-(3,4-Methylenedioxybenzyl)butyl(methyl)amine;* (34) *1-(3,4-Methylenedioxybenzyl)butyl(ethyl)amine*

(35) α-*Methyl-4-(methylthio)phenethylamine*

value in medicine with little abuse potential. Many of the latter are still of forensic interest because they metabolise to either amphetamine or methylamphetamine, which may be detected in the urine. Nearly all are N-substituted α-methylphenethylamines (*i.e.* N-substituted amphetamines). One of the simplest is methylamphetamine (Class B), the N-methyl derivative of amphetamine. Structure 7.4 shows the general form of an N-substituted amphetamine. Table 7.5 lists some of the better-known substances and a few examples of compounds which have been found in seizures. Derivatives where the amine nitrogen is part of a ring-structure (*e.g.* the Class C drug mesocarb) cannot be regarded as simple N-substituted amphetamines and are excluded from Table 7.5. The attraction of certain N-substituents to an illicit chemist is that a non-controlled drug can be made which is sufficiently labile that it can be converted metabolically or by other simple means into an active substance: in other words, it is a proxy for a controlled drug. A good example here is α-methylphenethylhydroxylamine (the N-hydroxy derivative of amphetamine), now listed as a Class B drug.

**Structure 7.4**   *The general structure of an* N-*substituted amphetamine*

### 7.4.2   Ring-substituted 'Cathinones'

Cathinone (a Class C drug; Structure 7.5) could form the basis of an equally large series of novel compounds. Whereas cathinone is found as a natural constituent of khat (see Chapter 9), the N-methyl homologue (methcathinone; Class B; Structure 7.6) is wholly synthetic. The N,N-diethyl derivative of cathinone is diethylpropion (amfepramone; Class C; Structure 7.7): a substance once used widely as an anorectic, but also abused for its stimulant properties. The N,N-di-t-butyl derivative of

**Table 7.5** *N-substituted amphetamines including established medicinal products, drugs of abuse and other illicit substances (see Structure 7.4)*

| Compound | R' | R'' | Comments |
|---|---|---|---|
| *N*-Acetylamphetamine | H | Acetyl | Illicit product |
| Amphetamine | H | H | Class B controlled drug |
| Amphetaminil | H | 1-Phenyl-1-cyanomethyl | Medicinal product (not UK) |
| Benzphetamine | H | Benzyl | Class C controlled drug |
| Clobenzorex | H | *o*-Chlorobenzyl | Medicinal product (not UK) |
| *N,N*-Dimethylamphet-amine | Methyl | Methyl | Illicit product |
| *N,N*-Di-(2-phenyliso-propyl)amine | H | 2-Phenylisopropyl | Illicit product |
| *N*-Ethylamphetamine | H | Ethyl | Class C controlled drug |
| Famprofazone | Methyl | 3-(1-Phenyl-2-methyl-4-isopropylpyrazolin-5-one)methyl | Medicinal product (not UK) |
| Fencamine | Methyl | (i) | Medicinal product (not UK) |
| Fenethylline | H | 7-Theophyllinylethyl[(ii)] | Class C controlled drug |
| Fenproporex | H | 2-Cyanoethyl | Class C controlled drug |
| Furfenorex | Methyl | 2-Furylmethyl | Medicinal product (not UK) |
| *N*-(2-Hydroxyethyl)-amphetamine | H | 2-Hydroxyethyl | Illicit product |
| Mefenorex | H | 3-Chloropropyl | Class C controlled drug |
| Methylamphetamine | H | Methyl | Class B controlled drug |
| α-Methylphenethyl-hydroxylamine | H | Hydroxy | Class B controlled drug |
| Prenylamine | H | 3,3-Diphenylpropyl | Medicinal product (not UK) |
| Selegiline | Methyl | 2-Propynyl | Medicinal product (UK) |

[(i)]The R'' substituent in fencamine is 3,7-dihydro-1,3,7-trimethyl-1*H*-purine-2,6-dione-8-amino-2-ethyl.
[(ii)]The term 'theophyllinyl' refers to 3,7-dihydro-1,3-dimethyl-1*H*-purine-2,6-dione.

cathinone is bupropion, an antidepressant drug used in the treatment of nicotine dependence. Ring-substituted 'cathinones' represent an almost totally unexplored family of compounds and their pharmacology is unknown.

**Structure 7.5**   *Cathinone*

**Structure 7.6**   *Methcathinone*

**Structure 7.7**   *Diethylpropion* (*amfepramone*)

### 7.4.3   Other Phenylalkyamines

In phenethylamine, the amino group is separated from the phenyl ring by two saturated carbon atoms. This configuration appears to be optimal for pharmacological activity. However, illicit chemists have experimented with other arrangements, none of which leads to controlled substances. In the mid-1990s, 1-phenylethylamine (α-methylbenzylamine; Structure 7.8) and, less commonly, the isomeric 4-methyl and *N*-methyl analogues (Structures 7.9 and 7.10 respectively) appeared in drug seizures in Europe. As far as is known, they behave as weak stimulants, but again, as with the 'cathinones', ring-substitution (particularly with alkoxy, alkylenedioxy or halogen) might offer substances with novel properties. A single example (1-phenyl-3-butanamine; also known as homoamphetamine; Structure 7.11) has been encountered where the amino group is more distant from the phenyl group.

**Structure 7.8**   *1-Phenylethylamine*

**Structure 7.9**   *4-Methyl-1-phenylethylamine*

**Structure 7.10**   *N-Methyl-1-phenylethylamine*

**Structure 7.11**   *1-Phenyl-3-butanamine*

## 7.5   OTHER POTENTIAL DESIGNER DRUGS

The substance *N*-benzylpiperazine (Structure 7.12), which may be regarded as a derivative of 1-phenylethylamine, together with certain analogues has been seen in drug seizures in the US and Europe.

**Structure 7.12**   *N-Benzylpiperazine*

Derivatives of the plant-based drugs, particularly cannabinols and cocaine have not been widely evaluated by clandestine chemists. This may partly reflect the wide availability of the parent product, but also because they are more complex structures than most phenethyl-amines. However, the finding of an illicit synthetic cocaine laboratory in

Spain in 2001 opens up the possibility that designer drugs based on natural products could appear. If the means are available to synthesise cocaine, then it would be particularly easy to create a series with varied substitution patterns in the phenyl ring. As mentioned in Chapter 3, the only known illicit derivative of a cannabinol is the acetyl ester of tetrahydrocannabinol.

In conclusion, there is still considerable scope for creating new non-controlled designer drugs from less common precursors. Some might have unexpected pharmacological properties.

*Chapter 8*

# Miscellaneous Issues

## 8.1 DRUGS REMOVED, REINSTATED OR RECLASSIFIED

In the past thirty years, the scope of the Act has increased to encompass many new substances. Indeed, the generic definitions discussed in Chapter 5 theoretically cover an infinite number. Yet in all this time, only two drugs, namely propylhexedrine and prolintane, have been permanently removed from control. Dexamphetamine (Class B) was also removed as a named substance in 1985 (S.I. 1995), but only because by then it fell to control as a stereoisomer of amphetamine.

Of the hundreds of named substances in the Act, most are never encountered. One reason for their retention is that they are still listed in UN1961 or UN1971 and few substances have ever been removed from international control. A second reason lies closer to home and is well illustrated by the history of pemoline. This anorectic drug was originally included because of its potential for abuse, but licensed products containing pemoline were no longer available by the early 1970s in the UK. Pemoline was therefore removed from the Act in 1973. Within a few years, illicit manufacturers began to produce pemoline tablets. Eventually, pemoline was brought under the scope of Schedule IV of UN1971 and pemoline was reintroduced into the Act in 1989. Apart from pemoline, phentermine and fencamfamin were also removed only to be later reinstated (see Table 8.1). The moral of this episode is that it is probably safer to leave substances under control even when they cease to be an immediate social problem. Unless, in the unlikely event that the pharmaceutical industry should wish to reactivate 'old' drugs, their retention causes no problems.

Because of commitments to the UN, there is in any case only limited scope to remove further substances. Two obvious candidates stand out in

**Table 8.1** *Drugs that have been removed or reinstated in the Act*

| Substance | Removed | Reinstated |
|---|---|---|
| Fencamfamin[(i)] | 1973 (S.I. 771) | 1986 (S.I. 2230) |
| Pemoline[(i)] | 1973 (S.I. 771) | 1989 (S.I. 1340) |
| Phentermine[(i)] | 1973 (S.I. 771) | 1985 (S.I. 1995) |
| Prolintane | 1973 (S.I. 771) | N/A |
| Propylhexedrine | 1995 (S.I. 1966) | N/A |

[(i)]The first three above were originally in Class C and were reinstated in Class C.

the Act, namely chlorphentermine and mephentermine. They are both in Class C and were originally brought under UK control in the DPMA 1964. Neither is listed in UN1971 and both have long ceased to be available in medicinal products in the UK. However, as with pemoline, their removal could send a signal to clandestine manufacturers.

Two named substances have been transferred between Classes. Nicodicodine was moved from Class A to Class B in 1973 (S.I. 771) and methaqualone was moved from Class C to Class B in 1984 (S.I. 859).

## 8.2   DEFINITION OF CANNABIS

In Section 37 of the Act, an original definition was set out as:
'["cannabis" (except in the expression "cannabis resin") means the flowering or fruiting tops of any plant of the genus *Cannabis* from which the resin has not been extracted, by whatever name they may be designated].'

Problems soon arose and led to a number of contentious cases. A new definition was introduced by Section 52 of the Criminal Law Act (1977):
'["cannabis" (except in the expression "cannabis resin") means any plant of the genus *Cannabis* or any part of any such plant (by whatever name designated) except that it does not include cannabis resin or any of the following products after separation from the rest of the plant, namely:

(a)  mature stalk of any such plant,
(b)  fibre produced from mature stalk of any such plant, and
(c)  seed of any such plant;]'.

Unlike some earlier legislation, the Act does not specify female plants nor any particular species of *Cannabis*, thereby avoiding taxonomic debate on whether or not that genus is monospecific.

### 8.3   PRODUCTION OF CRACK COCAINE

Cocaine base is a white amorphous solid. When seen in the form of small lumps (rocks) it is known as crack, although this is a colloquial term without a clear scientific definition. Unlike cocaine hydrochloride, cocaine base can be smoked. Pure cocaine base can be crystallised as fine needles, but is never seen in this form. Cocaine, which includes crack, is a Class A controlled drug. In R-*v*-Russell (see Appendix 7), it was held that the production of crack from cocaine hydrochloride (and by implication any salt/base interconversion) is an act of production for the purposes of Section 4 of the Act.

### 8.4   MEDICINAL PRODUCTS

#### 8.4.1   Introduction

The Class C anabolic steroids are subject to the provisions of Part II of Schedule 4 of the Regulations. They are excepted from the prohibition on possession as well as exportation and importation for personal use when in the form of a medicinal product. The original definition of a medicinal product is set out in Section 130 of the Medicines Act 1968. This has now been superseded by Article 1 of EEC Directive 65/65 as later implemented. The European definition states that a medicinal product is:

- 'Any substance or combination of substances presented for treating or preventing disease in human beings or animals'.
- 'Any substance or combination of substances which may be administered to human beings or animals with a view to making a diagnosis or to restoring, correcting or modifying physiological functions in human beings or animals is likewise considered a medicinal product'.

Many of the problems associated with controlled drugs in the form of medicinal product arose with the benzodiazepines. Schedule 4 of the Misuse of Drugs Regulations 1985 had excepted the benzodiazepines from the prohibition on possession when in the form of a medicinal product. However, this exception has been removed by the modified Part I of Schedule 4 of the Misuse of Drugs Regulations 2001 (see Appendix 4).

A number of uncertainties remain and the following situations have given rise to some debate.

### 8.4.2   Illicitly-made Tablets, Capsules or Injection Ampoules Containing a Class C Anabolic Steroid

These would probably be recognised as medicinal products since the definition is not dependent on whether the manufacturer was or was not licensed to make the products.

### 8.4.3   Crushed Tablets

Although the pure chemical substance used as the raw material for the manufacture of tablets is not a medicinal product, the situation with crushed tablets may depend on the circumstances of an individual case.

### 8.4.4   Abuse of Medicinal Products Containing Anabolic Steroids

A view has been proposed that steroid abusers are not taking the drug for medicinal purposes, especially if the products are intended for veterinary use only, and therefore those steroids are no longer a medicinal product.

## 8.5   'LOW-DOSAGE' PREPARATIONS

The purpose of Schedule 5 of the 2001 Regulations is to except a number of defined 'low-dose' preparations from certain controls. The full text is set out in Appendix 5. Although designed to remove onerous restrictions on the medical and pharmaceutical professions, some aspects of Schedule 5 can prove troublesome for forensic scientists.

The first problem arises with the requirements of paragraph 1.(1), particularly in relation to dihydrocodeine. In the UK, proprietary preparations containing dihydrocodeine include DHC Continus. This is available in 60 mg, 90 mg and 120 mg tablets. Because the dihydrocodeine is present as the bitartrate salt, even the 120 mg preparation contains only 80 mg of base; these tablets are therefore excepted by the provisions of paragraph 1.(1). But other dihydrocodeine tablets may be encountered where it is not necessarily obvious that the drug is present as the tartrate or any other specific salt. Since the origins of paragraph 1.(1) lie in UN1961, it is unlikely that manufacturers outside the UK would wish to market dihydrocodeine preparations containing more than 100 mg base, but if in doubt a quantitative analysis would have to be carried out.

The second area of difficulty concerns paragraph 2. Whilst almost all cocaine seized by police and customs will have a drug content well above 0.1%, this must be demonstrated in every case. For many purposes, the strength of the analytical signal, be that an IR spectrum or the total ion current from a GCMS, will show that this requirement has been met, but in other cases a quantitative analysis will be needed. It is likely that the 0.1% limit will eventually be removed; its continued inclusion has almost no benefit since pharmacies dispensing dilute cocaine solutions for hospital use will continue to prepare these from concentrated stock solutions to which the controls of Schedule 2 of the Regulations apply.

The exemption of morphine and opium preparations containing less than 0.2% morphine base is not usually problematic. Apart from proprietary morphine–kaolin mixtures, which always have less than 0.2% morphine, the only other common preparations are morphine sulfate tablets. These always contain much more than 0.2% morphine. It should be noted that, in paragraph 3 of Schedule 5, the term 'preparation of ...morphine' does not include derivatives of morphine. In R-*v*-Karagozlu (Inner London Crown Court, 2 December 1998), the defence argued successfully, but wrongly, that since the forensic analyst had not shown that the *diamorphine* concentration of a heroin exhibit was greater than 0.2%, there was no case to answer.

In some circumstances, the exceptions for low-dose forms do not apply. Paragraph 9 of Schedule 5 to the Regulations requires that in a mixture of one or more of the substances covered by paragraphs 1 to 8, no other controlled drug may be present. In other words, and assuming that 'controlled drug' here means a substance not already specified in Schedule 5, then if a mixture of amphetamine and cocaine, for example, were to be encountered, it is no longer necessary to show that the cocaine concentration exceeds 0.1%.

Apart from the exceptions listed in Schedule 5, there are no lower limits to the quantities of controlled drug required to create a possession offence. It is only necessary for the analyst to show unequivocally that the controlled drug is present. Legal arguments about what constitutes a usable amount of drug have largely been replaced by a need for the prosecution to show that the defendant was aware that he or she had possession of the drug.

## 8.6  DIAGNOSTIC KITS

In Regulation 2, certain products are exempted from controls. This includes so-called 'diagnostic kits', which may contain small amounts of controlled drugs. Such kits must first satisfy the requirements that they are not designed for administration of the drugs to a human or animal and that the drugs are not readily recoverable in a yield which constitutes a risk to health. They are then exempt provided that no one component part contains more than 1 mg of a controlled drug or 1 μg in the case of lysergide or any other *N*-alkyl derivative of lysergamide.

These threshold levels were designed to reduce the regulatory burden on manufacturers without opening an avenue for drug diversion. However, in framing the Regulation, some compromises were necessary; there are several substances where the required human dose is less than 1 mg, such that a kit containing these drugs could be misused. Although a number of rarely-seen narcotic analgesics fall into this category, a less unusual substance is DOB (4-bromo-2,5-dimethoxy-α-methyl phenethyl-amine), the effective dose of which is around 1 mg.

## 8.7  *PSILOCYBE* MUSHROOMS

The liberty cap mushroom *Psilocybe semilanceata* and certain other members of the *Psilocybe* and other fungal genera (*e.g. Amanita*) contain the hallucinogens psilocin (4-hydroxy-*N,N*-dimethyltryptamine) and its phosphate ester psilocybin, both of which are Class A drugs. Neither the cultivation nor possession of such mushrooms constitutes an offence under the Act. However, if the mushrooms are treated before use or preserved in some way (*e.g.* deliberate drying, cooking, freezing) then according to the judgement in R-*v*-Stevens and later cases (see Appendix 7), those processes are to be regarded as production of a Class A controlled drug.

## 8.8  MESCALINE

The peyote cactus (*Lophophora williamsii*) contains the hallucinogenic Class A controlled drug mescaline (3,4,5-trimethoxyphenethylamine). As with *Psilocybe* mushrooms, there is no possession offence in relation to the intact plant. Quite often, the separated outgrowths on the peyote cactus (mescal buttons) are seen, but there appears to have been no test case to determine whether these or any other deliberate treatment of peyote might constitute production of a Class A controlled drug.

## 8.9  POPPY-STRAW AND CONCENTRATE OF POPPY-STRAW

Part IV of Schedule 2 to the Act defines these as follows:

- '["poppy-straw" means all parts, except the seeds, of the opium poppy, after mowing]'.

- '["concentrate of poppy-straw" means the material produced when poppy-straw has entered into a process for the concentration of its alkaloids]'.

The dried seed head of an opium poppy, *i.e.* poppy-straw, often used in floristry, qualifies as a Class A drug, but by virtue of Regulation 4, it is exempt from most controls relating to possession, production or supply.

There are no special exceptions for concentrate of poppy-straw. Not only is this material rarely seen, but it is not always clear analytically how it differs from some forms of opium. Uncertainties also exist about what constitutes opium; these are discussed in Chapter 9.

## 8.10 PREPARATIONS DESIGNED FOR ADMINISTRATION BY INJECTION

Paragraph 6 of Part I of Schedule 2 to the Act extends control to: 'Any preparation designed for administration by injection specified in any of paragraphs 1 to 3 of Part II of this Schedule.' In practice, amphetamine is the only commonly-seen substance which might qualify under this definition. In other words, does a syringe containing a solution of amphetamine cause that drug to be treated as a Class A rather than a Class B substance? There is no clear answer to this question; it appears that the issue is rarely prosecuted and has not been seriously tested in a trial. One view is that the above definition was inserted into the Act following abuse by injection of commercially-produced methylamphetamine (*e.g.* Methedrine) in the 1960s. It is less certain that paragraph 6 was meant to apply to *ad hoc* solutions of illicitly-produced Class B drugs.

## 8.11 BENZODIAZEPINES

Until 2001, all but two of the benzodiazepines listed as Class C drugs were subject to less stringent controls. In particular, and apart from temazepam and flunitrazepam, there was no possession offence when in the form of a medicinal product. However, in 2001, Schedule 4 of the Regulations (see Appendix 4) was modified; there is now a possession offence for all benzodiazepines and increased controls on import and export. Temazepam and flunitrazepam remain in Schedule 3 of the Regulations.

## 8.12   DRUG 'INTERMEDIATES'

Not all substances listed in the Act are abusable as such; there are several examples of drug precursors/intermediates. They are all in Class A and all derive from UN1961 (see Table 8.2).

**Table 8.2** *Drug 'intermediates' listed in the Act*

| Substance | Alternative Name |
|---|---|
| 4-Cyano-2-dimethylamino-4,4-diphenylbutane | Methadone intermediate |
| 4-Cyano-1-methyl-4-phenylpiperidine | Pethidine intermediate A |
| Ecgonine | Cocaine precursor |
| 2-Methyl-3-morpholino-1,1-diphenylpropanecarboxylic acid | Moramide intermediate |
| 1-Methyl-4-phenylpiperidine-4-carboxylic acid | Pethidine intermediate C |
| 4-Phenylpiperidine-4-carboxylic acid ethyl ester | Pethidine intermediate B |

*Chapter 9*

# Future Developments

## 9.1 ISOTOPIC VARIANTS

### 9.1.1 Chemical Background

Each of the chemical elements (*e.g.* hydrogen, carbon, oxygen) exists in a number of isotopic forms. These isotopes may be stable or unstable (radioactive). The isotopes of an element arise from the presence of different numbers of neutrons in their atomic nucleus. Their electronic structure and qualitative chemical properties are unchanged, but because they differ in mass, slight differences exist between the quantitative properties of the isotopes of a given element. Both stable and radioactive isotopes are widely used in analytical-chemical, diagnostic and other medical procedures.

The lightest element is hydrogen and it has three isotopes. The most abundant form (99.985%) is known simply as hydrogen or protium ($^1$H). The nucleus contains one proton. Deuterium ($^2$H or D) is also a stable isotope of hydrogen; the nucleus contains a proton and a neutron. The natural abundance of deuterium is 0.015%. The nucleus of a third isotope, known as tritium ($^3$H or T) contains two neutrons and is radioactive. Tritium has an extremely low abundance, and is normally manufactured in nuclear reactors. The enrichment of a chemical compound so as to increase the proportion of D is called deuteration. Ordinary carbon consists of two stable isotopes: 98.9% of mass twelve ($^{12}$C) and 1.1% of mass thirteen ($^{13}$C). Apart from hydrogen, isotopes of other elements do not have unique names or symbols.

### 9.1.2 A Case History

In a criminal trial in Sweden in the late 1990s, the defendants were charged with the unlawful manufacture of amphetamine. Their defence was that they intended to produce deuterated amphetamine, which was not a scheduled substance. After much debate and conflicting expert advice, the Supreme Court in Stockholm decided by a majority verdict on 5 July 1999 that deuterated amphetamine was to be considered as a substance under the Swedish Penal Law on Narcotics.

There is now an almost unanimous view that the specification of a particular isotope which forms part of a controlled substance does not influence the fact that the substance is subject to control. If practical matters are ignored, *i.e.* whether suitable precursors are available and if the process is economically attractive, then the main arguments in favour of this view are as follows:

- Isotopically-pure substances do not exist. In the case of normal amphetamine, where there are nine carbon and thirteen hydrogen atoms, 0.2% of molecules will have at least one hydrogen replaced by deuterium and 10% of molecules will have at least one $^{13}C$ atom. An extreme example of this is afforded by the Class A controlled drug 4-bromo-2,5-dimethoxy-α-methylphenethylamine (DOB). Considering the bromine atom in this molecule and ignoring isotopic variation in other atoms, then DOB exists in two almost equally abundant forms, *i.e.* a form containing $^{79}Br$ (50.6%) and a form containing $^{81}Br$ (49.4%). It can hardly be argued that one is controlled and the other not; the Act must cover both. If DOB were to be enriched or depleted in either bromine isotope, the product would still be controlled. By extension, the same argument applies to all isotopes.
- The biological properties of isotopic variants of controlled drugs differ only slightly from the normal compounds. There is no evidence that they cause any less social harm.
- Isotopic variants are not distinct chemical entities.

A dissenting view is based on a conservative interpretation of the UN Conventions, namely there is no explicit mention of these variants in the international drug control treaties.

### 9.1.3   Implications for the Misuse of Drugs Act

There remains a possibility that any future case involving isotopic variants could again lead to lengthy technical and legal arguments in a court. If it were felt that, despite the above arguments, the status of isotopic variants should be clarified unambiguously in the Act, then new paragraphs could be added to Parts I, II and III of Schedule 2 referring to 'Any isotopic variants of a substance for the time being specified in paragraphs 1 or 2 (*etc.*) of this Part of this Schedule.'

## 9.2   ANABOLIC STEROIDS – OTHER ISSUES

### 9.2.1   Further Substances for Control

In 1996, forty-eight anabolic/androgenic steroids together with some generic definitions were added to the Act. The substances are essentially those proscribed by the International Olympic Committee (IOC). Further steroids were recently added to the IOC list and it is expected that the following four will be added by name to the Act as Class C drugs; none falls within the scope of the generic definition:

- 4-androstene-3,17-dione
- 19-nor-4-androstene-3,17-dione
- 5-androstene-3,17-diol
- 19-nor-5-androstene-3,17-diol

### 9.2.2   Supply of Meat Products

Anabolic/androgenic steroids are derivatives of testosterone, which is itself controlled. Thus for the first time, the Act includes a compound that occurs naturally in the tissues of both male and female humans, other mammals and birds. While possession of anabolic/androgenic steroids is not an offence, their unauthorised supply is illegal. This then raises a legal problem. If testosterone is present in human blood (the level is typically 5–10 micrograms per litre), then providing transfusion blood could be deemed to be supply. In reality this issue is not new, since who can be sure that transfusion blood never contains any other controlled substance? This is in any case a somewhat academic point since a prosecution is not likely to be in the public interest. But a more difficult question arises with the supply of meat products; they also contain testosterone. The levels may be low, but the Act makes no current provision for a *de minimis* approach to steroids.

## 9.3   HASH OIL

Cannabis (hash) oil has traditionally been made by solvent extraction of cannabis resin followed by removal of the excess solvent to leave a dark viscous liquid. It may contain ten times as much tetrahydrocannabinol (THC) as cannabis and cannabis resin.

Preparations or products of Class B drugs are also controlled by virtue of paragraph 4 of Part II of Schedule 2, namely 'Any preparation or other product containing a substance or product for the time being specified in any of paragraphs 1 to 3 of this Part of this Schedule, not being a preparation falling within paragraph 6 of Part I of this Schedule'. Hash oil is not considered to fall within this definition because it cannot be said to contain cannabis or cannabis resin as such. However, Section 37 of the Act defines cannabis resin as '...the separated resin, whether crude or purified, obtained from any plant of the genus *Cannabis*'. It has been accepted that hash oil is a purified form of cannabis resin and therefore a Class B drug.

This situation was brought into confusion after 1990 when it became clear that some hash oil was being produced, not from cannabis resin, but from herbal cannabis. In R-*v*-Carter (Oxford Crown Court, Judgement of 16 December 1992) it was successfully argued that such hash oil could no longer be deemed to be a purified form of cannabis resin; the only option open to the court was to regard it as a preparation of cannabinol and therefore a Class A drug.

It is sometimes possible to distinguish the two types of hash oil insofar as cannabidiol is present in much greater amounts in cannabis resin than it is in cannabis. Thus if cannabidiol is found in hash oil, then it has probably not originated from herbal cannabis. Hash oil made from herbal cannabis may also appear to have a dark green colour because of the presence of the pigment chlorophyll. Since, from the users viewpoint, there is little distinction between the two types of hash oil, it is anomalous that it could be treated as either Class A or Class B solely on the basis of its manufacturing route.

Following the above judgement, there arose a general scientific agreement that a solution to the problem would be to define hash oil as a Class B drug by including it as a named substance in Part II of Schedule 2 to the Act. Unfortunately, unlike chemically-defined substances, hash oil as an entity cannot be added to the Act by a simple Modification Order. The reason for this is that firstly a definition of hash oil would have to be incorporated into Part IV of Schedule 2 to the Act if not also into Section 37. Secondly, the definition of the Class A cannabinols in Part IV of Schedule 2 would have to be re-worded to say 'cannabinol derivatives means the following substances, except where contained in cannabis, cannabis resin *or liquid cannabis....*'. A change to Section 37 would require primary legislation. A suitable definition might be 'Liquid cannabis is a purified form/solvent extract of cannabis or cannabis resin'. The term 'liquid cannabis' has some advantages over 'hash oil' or 'cannabis oil' since it does not beg the question of the definition of an oil. The use of 'purified form' rather than 'solvent extract' may be advisable because the latter might require the forensic scientist to prove that a solvent was used. It is interesting to note that UN1961 included the concept of liquid cannabis, but defined it as '. . .extracts and tinctures of cannabis'. This was a reference to the medicinal products once found in pharmacopoeias, but now long obsolete.

If cannabis and cannabis resin were to be re-classified as Class C drugs (see later) then, in principle, hash oil could follow them into Class C. As noted earlier, problems have been caused by the separation in the Act between the Class A cannabinols and the corresponding Class B plant material. In fact this is the only instance in the Act where the classification of a pharmacologically-active substance (*i.e.* THC) is effectively based on the potency of different products or preparations. Hash oil forms an awkward bridge between these two groups. If the separation of cannabinols and plant material were widened further (*i.e.* Class A and Class C), then the situation with hash oil would lead to an even greater anomaly. This could be solved by ensuring that all cannabis related substances have the same classification.

## 9.4   OPIUM

### 9.4.1   Introduction

The different forms of opium are mentioned at various places in the Act:

- Paragraph 1 of Part I of Schedule 2 specifies 'Opium, whether raw, prepared or medicinal' as being a Class A drug.
- Paragraph 5 of the same Part extends control to include 'any preparation or product containing [opium]'.
- In Part IV of Schedule 2, there are two further definitions:

  (i) '["medicinal opium" means raw opium which has undergone the process necessary to adapt it for medicinal use in accordance with the requirements of the British Pharmacopoeia, whether it is in the form of powder or is granulated or is in any other form, and whether it is or is not mixed with neutral substances;]'
  (ii) '["raw opium" includes powdered or granulated opium but does not include medicinal opium.]'

- Section 37(1) offers a further definition: '["prepared opium" means opium prepared for smoking and includes dross and any other residues remaining after opium has been smoked;]'

The Regulations also refer to opium in several ways:

- Schedule 1 lists 'raw opium', and Schedule 2 lists 'medicinal opium', but prepared opium is not included.
- Paragraph 3 of Schedule 5 makes an exception from the prohibition on importation, exportation and possession (subject to the requirements of Regulations 24 and 26) for certain 'low concentration' preparations of medicinal opium.
- A further exception is made in paragraph 8 of Schedule 5. This concerns mixtures of opium and ipecacuanha.

Finally, it should be noted that, apart from the general prohibitions on possession, possession with intent to supply *etc.* relating to all drugs in Schedule 2, there are specific offences within Sections 8 and 9 of the Act which refer to opium by name.

### 9.4.2   Definitions of Opium

The difficulty faced by the forensic scientist when dealing with suspected opium is that although raw, prepared and medicinal opium and opium preparations are apparently defined in the Act or Regulations, there is no clear statement of what constitutes opium. This problem is partly caused because opium is rarely seen in casework. As a consequence, analysts do not have the day-to-day familiarity that applies to, say, cannabis products. It is also clear that confusion exists as to the distinguishing features of the different forms of opium. A second level of difficulty arises because definitions do exist, but they are unsuitable for the forensic chemist. For example, in The British Pharmacopoeia (BP), raw opium is defined as '...the air-dried latex obtained by incision from the unripe capsules of *Papaver somniferum* L. It contains not less than 10.0 per cent morphine... and not less than 2.0 per cent codeine...'. The New Shorter Oxford English Dictionary defines opium as 'A reddish-brown strong-scented addictive drug prepared from the thickened dried juice of the unripe capsules of the opium poppy, used as a stimulant and intoxicant, and in medicine as a sedative and analgesic'.

Recourse to the BP definition is unacceptable because it is not technically possible to identify the botanical origin of opium. Moreover, the capsule exudates (latex) from certain other species of the *Papaver* genus also contain morphine and related alkaloids. Secondly, if a sample contains less than 10.0% morphine or less than 2.0% codeine then it is presumably not 'Opium of the Pharmacopoeia'. It is therefore inappropriate that the legal definition should devolve onto a BP definition primarily developed to ensure the quality of an item of trade. The dictionary definition is similarly inadequate in that it refers to the origin and pharmacological properties of opium rather than its chemical constitution.

### 9.4.3 Current Forensic Science Practice

In the absence of clear criteria, analysts opt to give an opinion as to whether a sample is opium based on their knowledge and experience of the typical appearance and smell of opium. It is usually possible to demonstrate that the substance is a preparation or product containing a controlled drug by identifying the morphine present. However, this would not prove that the sample is opium, so opium-specific offences will not apply.

### 9.4.4 Possible Modification to the Act

A definition of opium, which could be inserted into the Act, might be:
   'Opium is a resinous material containing a range of alkaloids including morphine and codeine.'
   However, even then it would not be entirely clear how an analyst would distinguish opium (made by capsule incision) from concentrate of poppy-straw (made by chemical extraction of poppy-straw), which is listed separately as a Class A drug in the Act.

## 9.5  GAMMA-HYDROXYBUTYRATE (GHB)

Gamma-hydroxybutyrate (4-hydroxybutyrate; GHB) acts as a central nervous system depressant, and is chemically related to the brain neurotransmitter gamma-aminobutyric acid (GABA). Synonyms of GHB include oxybate, gamma-OH, somatomax, 'GBH' and 'liquid ecstasy'. It is not a licensed medicine in the UK, but GHB is used clinically in some European countries as an anaesthetic drug. It is easily manufactured by adding aqueous sodium hydroxide to gamma-butyrolactone (GBL) to leave a weakly alkaline solution.

In early 2001, the UN decided that gamma-hydroxybutyrate (GHB) would be added to Schedule IV of UN1971. It is expected that the UK will list GHB as a Class C drug, subject to the controls of Schedule 4 Part I of the Regulations. Although the term 'gamma-hydroxybutyrate' includes both salts and the free acid, it would be appropriate to list GHB in the Act as 4-hydroxybutyric acid since its salts will be automatically subsumed. A major technical difficulty arising from the proposed control of GHB will be that the above chemical reaction is reversible as shown in Structures 9.1(1) to 9.1(3).

(1)                          (2)                                    (3)

**Structure 9.1**   *The interconversion of gamma-hydroxybutyrate (GHB) and its precursor gamma-butyrolactone (GBL). (1) 4-Butyrolactone (GBL); (2) 4-Hydroxybutyric acid (GHB); (3) 4-Hydroxybutyric acid sodium salt (GHB)*

The precursor (sometimes known by the synonym dihydrofuranone) can be simply recovered from a GHB solution by adding acid to neutralise the sodium hydroxide. It is only necessary to add a small amount of water to GBL to produce a detectable level of GHB. This interconversion occurs naturally in the body and drug users have realised that it is possible to ingest the precursor directly to produce the desired effect. Unfortunately, any attempt to control the precursor would prove difficult; gamma-butyrolactone is inexpensive and widely used as an industrial solvent.

## 9.6   KETAMINE

Ketamine is a dissociative anaesthetic drug used in animal and occasional emergency human surgery. It is not strictly hallucinogenic, but causes catalepsis (muscle rigidity) and leaves users feeling detached from their immediate environment. In the UK, ketamine is available for hospital use as injection solutions, but there are no preparations licensed for oral use.

Abuse of ketamine was recognised almost thirty years ago in the USA, but it did not come to notice until 1990 in the UK. At that time, seizures often comprised ampoules of the proprietary preparations Ketalar and Vetalar or loose powders which had probably been produced by evaporation of these injection liquids. During the mid- to late 1990s, nearly all illicit ketamine was found in the form of well-made tablets, visually similar to, and often sold as, 'Ecstasy' tablets. In the past year, loose powders have become more common. This may be a reflection of the illicit market responding to the several successful prosecutions of ketamine tablet manufacturers for attempted supply of a controlled drug, *i.e.* a MDMA 'look-alike'.

Ketamine is not controlled by the Act, but is closely related to phencyclidine (a Class A controlled drug) and to the (non-controlled) veterinary anaesthetic tiletamine (Structures 9.2 to 9.4). In 2000, a risk assessment was carried out on ketamine by the European Monitoring Centre for Drugs and Drug Addiction (EMCDDA; see Chapter 1), but there was insufficient evidence to recommend that it should be controlled by Member States of the European Union.

**Structure 9.2** *Ketamine*

**Structure 9.3** *Phencyclidine*

**Structure 9.4** *Tiletamine*

## 9.7 ZOLPIDEM

In early 2001, the UN CND meeting decided that a further four substances should be brought within the scope of UN1971, namely

4-MTA and 2C-B (both already controlled under the Act), GHB (see above) and zolpidem. It is likely that zolpidem will be treated in a similar way to most benzodiazepines (*i.e.* as a Class C drug subject to Part I of Schedule 4 of the Regulations). Although zolpidem is structurally unrelated to the benzodiazepines, its pharmacology and abuse potential are broadly similar.

## 9.8   CANNABIS SEEDS

Cannabis seeds are excluded from control, but despite 'sterilisation', even seeds intended for use as bird food or fishing bait will often germinate. A new offence involving sale of seeds intended for cannabis cultivation and related cultivation equipment has been suggested, but, as far as is known, this has not been progressed. It is conceivable that cannabis seeds could be regarded as precursors to cannabis and therefore incorporated into the appropriate legislation (see Appendix 6). But apart from the uses noted above, cannabis seeds of approved types are produced on a large scale for the licensed cultivation of 'low-THC' crops. These are intended for the manufacture of rope, paper and animal bedding; any further controls on seed would impact on this industry.

Although small amounts of cannabis (*e.g.* bracts and other flowering parts) may be found adhering to cannabis seeds, there are insignificant quantities of THC in clean seeds. However, some oils made from cannabis seeds, often intended for culinary or cosmetic purposes, do contain measurable THC levels. While not representing a realistic opportunity for abuse, it has been claimed that consumption of seed oil can lead to a positive result for cannabinoids when urine is screened by sensitive immunoassay techniques.

## 9.9   DIHYDROETORPHINE AND REMIFENTANIL

In 1999, the UN CND decided that these substances should be added to Schedule I of UN1961. Both are potent analgesics; the former is chemically closely related to the Class A drug etorphine. Remifentanil is an analogue of fentanyl, but is not subsumed by the generic definition in paragraph 1(d) of Part I of Schedule 2. It is likely that both dihydroetorphine and remifentanil will become Class A drugs subject to Schedule 2 of the Regulations.

## 9.10 OTHER CANDIDATES FOR CONTROL

### 9.10.1 Khat

Drug abuse is not restricted to substances in the Act. Apart from ketamine (see above), it is frequently suggested that khat might also be brought under control. Khat (also known as qat or chat) comprises the leaves and fresh shoots of *Catha edulis*, a flowering evergreen shrub cultivated in East Africa and the Arabian peninsula. The active components, cathinone [(*S*)-1-amino-propiophenone] and cathine [(*S*,*S*)-norpseudoephedrine], are usually present at around 0.3 to 2.0%. Both substances are close chemical relatives of synthetic drugs such as amphetamine and methcathinone. Khat is controlled in some European countries, the USA and Canada. Although *Catha edulis* is not under international control, cathine and cathinone are listed in UN1971 under Schedules III and I respectively; in the UK they are both Class C substances. Alcoholic extracts (tinctures) of khat have been noted especially in 'Herbal High' sales outlets and at music festivals. The only known prosecution under the Act for the unlawful production of cathine and cathinone from khat was unsuccessful (R-*v*-Farmer, Lewes Crown Court, 1998). Using thin-layer chromatography, cathine can be distinguished from its non-controlled diastereoisomers [(*S*,*R*)- and (*R*,*S*)-norephedrine (phenylpropanolamine); see Chapter 3].

### 9.10.2 Ephedrine

Other herbal materials may be submitted for forensic examination, but they are not common. Ephedrine, either as a synthetic substance or in the form of extracts of *Ephedra vulgaris* (also known as Ma Huang in Chinese medicine), is encountered either alone or mixed with other drugs such as ketamine in powders or tablets. Ephedrine and pseudoephedrine are recognised as substances useful for the manufacture of methylamphetamine and methcathinone; they are subject to certain trade controls under the provisions of UN1988 and subsequent domestic legislation (see Appendix 6). A few years ago, the WHO proposed that ephedrine should be brought within the scope of UN1971, which would have lead to it becoming a controlled drug in the UK. However, that proposal was not accepted by the majority of UN signatories.

### 9.10.3 Other Substances

Even though powdered seeds of morning glory (*Ipomoea* species) are part of the 'Herbal High' repertoire, no prosecutions have been forthcoming for offences involving the preparation of a controlled drug, namely lysergamide. *Salvia divinorum* contains a non-controlled hallucinogen (salvinorin A), but, as with many other obscure hallucinogens, it has not achieved any popularity amongst drug users.

Zopiclone, a prescription only hypnotic drug, is not otherwise under domestic or international control, but is already showing early signs of abuse and might be a candidate for future scheduling under UN1971. Dextromethorphan (normally used as an antitussive), nalbuphine (a prescription analgesic similar to other opiates) and volatile solvents have all been subject to abuse. The solvents comprise low molecular weight alkanes (*e.g.* butane from cigarette lighter fuels), toluene (a solvent in some glues) and various aerosol propellants. Even though these chemicals continue to be associated with fatal poisonings, particularly in young people, they are extremely ubiquitous and it is unlikely that they could ever be brought within the scope of the Act. However, there is legislation which makes it an offence to supply certain of these products to young people. Alkyl nitrites form a distinct sub-group of solvents. Although amyl nitrite has recognised value as a coronary vasodilator and antidote to cyanide poisoning, illicit products (so-called 'poppers') often contain isobutyl nitrite. Alkyl nitrites are unlikely subjects for control under the Act, but legal proceedings are in hand to determine if they could constitute a medicinal product under the Medicines Act.

## 9.11   RECLASSIFICATION OF CANNABIS

In late 2001, the Home Secretary announced that the Government was referring the status of cannabis and cannabis resin to the Advisory Council on the Misuse of Drugs. Reclassification of cannabis and cannabis resin from Class B to Class C and cannabinols from Class A to Class C was supported by the Advisory Council. This recommendation had originally been made in the Report of the Independent Enquiry into the Misuse of Drugs Act (see Bibliography). There is no intention that cannabis, cannabis resin or cannabinols should be rescheduled with respect to the Regulations unless there is evidence from current research that cannabis has medicinal uses.

## 9.12   RECLASSIFICATION OF OTHER DRUGS

Apart from cannabis, cannabis resin and the cannabinols, the Independent Enquiry into the Misuse of Drugs Act also recommended that a number of other controlled drugs should be reclassified. It proposed that 'Ecstasy' and related compounds should be moved from

Class A to B, LSD from Class A to B, and buprenorphine from Class C to B. The Government rejected some of these recommendations. However, they are likely to be considered again by the Home Affairs Select Committee during its current review of the UK drugs strategy. One of the outstanding difficulties concerning any debate on the re-classification of MDMA is the uncertainty about its neurotoxic effects in humans.

# Bibliography

R. Fortson, *Misuse of Drugs and Drug Trafficking Offences*, 4th edition, Sweet and Maxwell, London, 2002.

L. Jason-Lloyd, *Drugs, Addiction and the Law*, 7th edition, ELM Publications, Huntingdon, 2002.

J.M. Corkery, *Drug Seizure and Offender Statistics*, Home Office Research, Development and Statistics Directorate, London, 2001.

M. Ramsay, P. Baker, C. Goulden, C. Sharp and A. Sondhi, *Drug Misuse Declared in 2000: Results from the British Crime Survey*, Home Office Research Study No. 224, Home Office Research, Development and Statistics Directorate, London, 2001.

The Police Foundation, *Drugs and the Law: Report of the Independent Inquiry into the Misuse of Drugs Act 1971*, London, 2000.

Advisory Council on the Misuse of Drugs, *Drug Misuse in the Environment*, The Stationery Office, London, 1998.

P. Bucknell and H. Ghodse, *Bucknell and Ghodse on Misuse of Drugs*, 3rd edition, Sweet and Maxwell, London, 1997.

A. Shulgin and A. Shulgin, *TIHKAL, The Continuation*, Transform Press, Berkeley, California, 1997.

Parliamentary Office of Science and Technology, *Common Illegal Drugs and Their Effects*, House of Commons, London, 1996.

S.B. Karch, *The Pathology of Drug Abuse*, 2nd edition, CRC Press, New York, 1996.

J. Hardman (ed.), *Goodman and Gilman's Pharmacological Basis of Therapeutics*, 9th edition, McGrawHill, New York, 1996.

United Nations International Drug Control Programme, *Multilingual Dictionary of Narcotic Drugs and Psychotropic Substances Under International Control*, United Nations, New York, 1993.

A. Shulgin and A. Shulgin, *PIHKAL: A Chemical Love Story*, Transform Press, Berkeley, California, 1992.

T.A. Gough (ed.), *The Analysis of Drugs of Abuse*, John Wiley and Sons, New York, 1991.

M. Klein, F. Sapienza, H. McClain and I. Khan (eds), *Clandestinely Produced Drugs, Analogues and Precursors: Problems and Solutions*, United States Department of Justice Drug Enforcement Administration, Washington, DC, 1989.

*Appendix 1*

# Modification Orders to the Misuse of Drugs Act 1971

*A.1.1   The Misuse of Drugs Act 1971 (Modification) Order 1973*
*(S.I. 771)*

Transfers nicodicodine from Part I to Part II of Schedule 2 and excludes from Part I certain substances (notably codeine, dihydrocodeine, ethyl-morphine, norcodeine and pholcodeine) which are already included in Part II. The Order also adds drotebanol to Part I and propiram to Part II and removes from Part III fencamfamin, pemoline, phentermine and prolintane.

*A.1.2   The Misuse of Drugs Act 1971 (Modification) Order 1975*
*(S.I. 421)*

Adds difenoxin and 4-bromo-2,5-dimethoxy-α-methylphenethylamine to Part I of Schedule 2.

*A.1.3   The Misuse of Drugs Act 1971 (Modification) Order 1977*
*(S.I. 1243)*

Adds certain tryptamine derivatives and certain phenethylamine derivatives to Part I of Schedule 2.

*A.1.4   The Misuse of Drugs Act 1971 (Modification) Order 1979*
*(S.I. 299)*

Adds phencyclidine to Part I of Schedule 2.

### A.1.5   The Misuse of Drugs Act 1971 (Modification) Order 1983
### (S.I. 765)

Adds sufentanil and tilidate to Part I of Schedule 2 and dextropropoxy-phene to Part III.


### A.1.6   The Misuse of Drugs Act 1971 (Modification) Order 1984
### (S.I. 859)

Adds alfentanil, eticyclidine, rolicyclidine and tenocyclidine to Part I of Schedule 2, certain barbiturates (that is to say 5,5-disubstituted barbituric acids and methylphenobarbitone) and mecloqualone to Part II, transfers methaqualone from Part III to Part II and adds diethylpropion to Part III.


### A.1.7   The Misuse of Drugs Act 1971 (Modification) Order 1985
### (S.I. 1995)

Adds glutethimide, lefetamine and pentazocine to Part II of Schedule 2, removes explicit reference to dexamphetamine, and adds ethchlorvynol, ethinamate, mazindol, meprobamate, methyprylone, phentermine and a group of 33 benzodiazepines to Part III.


### A.1.8   The Misuse of Drugs Act 1971 (Modification) Order 1986
### (S.I. 2230)

Adds carfentanil, lofentanil, certain fentanyl derivatives and certain pethidine derivatives to Part I of Schedule 2 and cathine, cathinone, fencamfamin, fenethylline, fenproporex, mefenorex, propylhexedrine, pyrovalerone and N-ethylamphetamine to Part III.


### A.1.9   The Misuse of Drugs Act 1971 (Modification) Order 1989
### (S.I. 1340)

Adds buprenorphine and pemoline to Part III of Schedule 2.


### A.1.10   The Misuse of Drugs Act 1971 (Modification) Order 1990
### (S.I. 2589)

Adds N-hydroxy-tenamphetamine and 4-methyl-aminorex to Part I of Schedule 2 and midazolam to Part III.

### A.1.11   The Misuse of Drugs Act 1971 (Modification) Order 1995 (S.I. 1966)

Removes propylhexedrine from Part III of Schedule 2.

### A.1.12   The Misuse of Drugs Act 1971 (Modification) Order 1996 (S.I. 1300)

Adds certain anabolic/androgenic steroids, clenbuterol and certain polypeptide hormones to Part III of Schedule 2.

### A.1.13   The Misuse of Drugs Act 1971 (Modification) Order 1998 (S.I. 750)

Adds etryptamine to Part I of Schedule 2, methcathinone and zipeprol to Part II and aminorex, brotizolam and mesocarb to Part III.

### A.1.14   The Misuse of Drugs Act 1971 (Modification) Order 2001 (S.I. 3932)

Adds 35 phenethylamine derivatives to Part I of Schedule 2 and α-methylphenethylhydroxylamine to Part II.*

*S.I. 3932 came into force on 1st February 2002.

*Appendix 2*

# Part IV of Schedule 2 to the Misuse of Drugs Act 1971

The following is the full text of Part IV.

MEANING OF CERTAIN EXPRESSIONS USED IN THIS SCHEDULE

For the purposes of this Schedule the following expressions (which are not among those defined in Section 37 (1) of this Act) have the meanings hereby assigned to them respectively, that is to say:

'Cannabinol derivatives' means the following substances, except where contained in cannabis or cannabis resin, namely tetrahydro derivatives of cannabinol and 3-alkyl homologues of cannabinol or its tetrahydro derivatives;

'Coca leaf' means the leaf of any plant of the genus *Erythroxylon* from whose leaves cocaine can be extracted either directly or by chemical transformation;

'Concentrate of poppy-straw' means the material produced when poppy-straw has entered into a process for the concentration of its alkaloids;

'Medicinal opium' means raw opium which has undergone the process necessary to adapt it for medicinal use in accordance with the requirements of the British Pharmacopoeia, whether it is in the form of powder or is granulated or is in any other form, and whether it is or is not mixed with neutral substances;

'Opium poppy' means the plant of the species *Papaver somniferum* L;

'Poppy-straw' means all parts, except the seeds, of the opium poppy, after mowing;

'Raw opium' includes powdered or granulated opium but does not include medicinal opium.

# Overview of Schedules 1 to 5 of the Misuse of Drugs Regulations 2001 (S.I. 3998)

The Regulations of 2001 came into force on 1st February 2002. The organisation of controlled drugs into five Schedules in the Regulations is based on a balance between their value as medicines and their hazards as drugs of abuse. In broad terms, at least for psychotropic drugs, the Schedules correspond to the respective Schedules of the 1971 Convention. Whereas it is permitted for a substance listed in UN1971 to be placed in a higher (more stringent) Schedule in the Regulations, it may not be placed lower. The following are listed in Schedule IV of UN1971, but Schedule 3 in the Regulations: benzphetamine, diethylpropion, ethchlorvynol, ethinamate, mazindol, meprobamate, methylphenobarbitone, methyprylone, phendimetrazine, phentermine, pipradrol and temazepam. In addition, lefetamine is in Schedule IV of UN1971, but Schedule 2 in the Regulations and glutethimide is III and 2 respectively. The connection between the Regulations and the Schedules of UN1961 is less precise, although the principle still holds that national governments must not permit less stringent controls on substances than those set out in UN1961.

Controls are placed on the manufacture, prescription and record keeping of the substances in decreasing order 1 to 5. Drugs in Schedule 1 may not be prescribed, but can be used under licence in medical and scientific research, whereas substances in Schedule 4 Part II, provided they are in the form of a medicinal product, are freely available to the extent that there is no possession offence for 'personal use'. Further exceptions to certain offences with some 'low-dose' preparations occur in Schedule 5. There is much less correlation between the assignment of a substance in the Act and its occurrence in a Schedule of the Regulations. Table A3.1 gives examples of the two-dimensional matrix of UK drug control. Most Class C drugs are found in Schedule 4 and most Class A drugs are found in Schedules 1 and 2 of the Regulations.

Forensic scientists are mostly concerned with the requirements of Schedules 4 and 5 (see Appendix 4 and Appendix 5). A cross-reference to the Schedule in the Regulations and the Class in the Act for all controlled drugs can be found in Tables 2.1 to 2.3 in Chapter 2.

**Table A3.1** *The relationship between Class in the Act and Schedule in the Regulations for selected substances*

| Regulations | Class A | Class B | Class C |
|---|---|---|---|
| Schedule 1 | Lysergide, MDMA | Cannabis, methcathinone | Cathinone |
| Schedule 2 | Diamorphine, cocaine | Amphetamine, codeine | Dextropropoxyphene |
| Schedule 3 | – | Barbiturates (most) | Temazepam, flunitrazepam |
| Schedule 4 Part I | – | – | Benzodiazepines (most) |
| Schedule 4 Part II | – | – | Anabolic steroids |
| Schedule 5 | 'Low dose' cocaine | 'Low dose' codeine | 'Low dose' dextropropoxyphene |

*Appendix 4*

# Schedule 4 to the Misuse of Drugs Regulations 2001

The following is the full text of Schedule 4 as it appears in the Regulations.

## Part I

Controlled Drugs Subject to the Requirements of Regulations 22, 23, 26 and 27:

1.

| | |
|---|---|
| Alprazolam | Haloxazolam |
| Aminorex | Ketazolam |
| Bromazepam | Loprazolam |
| Brotizolam | Lorazepam |
| Camazepam | Lormetazepam |
| Chlordiazepoxide | Medazepam |
| Clobazam | Mefenorex |
| Clonazepam | Mesocarb |
| Clorazepic acid | Midazolam |
| Clotiazepam | Nimetazepam |
| Cloxazolam | Nitrazepam |
| Delorazepam | Nordazepam |
| Diazepam | Oxazepam |
| Estazolam | Oxazolam |
| N-Ethylamphetamine | Pemoline |
| Ethyl loflazepate | Pinazepam |
| Fencamfamin | Prazepam |
| Fenproporex | Pyrovalerone |
| Fludiazepam | Tetrazepam |
| Flurazepam | Triazolam |
| Halazepam | |

93

2. Any stereoisomeric form of a substance specified in paragraph 1.
3. Any salt of a substance specified in paragraph 1 or 2.
4. Any preparation or other product containing a substance or product specified in any of paragraphs 1 to 3, not being a preparation specified in Schedule 5.

## Part II

Controlled drugs excepted from the prohibition on possession when in the form of a medicinal product; excluded from the application of offences arising from the prohibition on importation and exportation when imported or exported in the form of a medicinal product by any person for administration to himself; and subject to the requirements of Regulations 22, 23, 26 and 27:

1. The following substances, namely:

| | |
|---|---|
| Atamestane | Methenolone |
| Bolandiol | Methyltestosterone |
| Bolasterone | Metribolone |
| Bolazine | Mibolerone |
| Boldenone | Nandrolone |
| Bolenol | Norboletone |
| Bolmantalate | Norclostebol |
| Calusterone | Norethandrolone |
| 4-Chloromethandienone | Ovandrotone |
| Clostebol | Oxabolone |
| Drostanolone | Oxandrolone |
| Enestebol | Oxymesterone |
| Epitiostanol | Oxymetholone |
| Ethyloestrenol | Prasterone |
| Fluoxymesterone | Propetandrol |
| Formebolone | Quinbolone |
| Furazabol | Roxibolone |
| Mebolazine | Silandrone |
| Mepitiostane | Stanolone |
| Mesabolone | Stanozolol |
| Mestanolone | Stenbolone |
| Mesterolone | Testosterone |
| Methandienone | Thiomesterone |
| Methandriol | Trenbolone |

2. Any compound (not being Trilostane or a compound for the time being specified in paragraph 1 of this Part of this Schedule)

structurally derived from 17-hydroxyandrostan-3-one or from 17-hydroxyestran-3-one by modification in any of the following ways, that is to say:

(i) by further substitution at position 17 by a methyl or ethyl group;

(ii) by substitution to any extent at one or more of the positions 1, 2, 4, 6, 7, 9, 11 or 16, but at no other position;

(iii) by unsaturation in the carbocyclic ring system to any extent, provided that there are no more than two ethylenic bonds in any one carbocyclic ring;

(iv) by fusion of ring A with a heterocyclic system.

3. Any substance which is an ester or ether (or, where more than one hydroxyl function is available, both an ester and an ether) of a substance specified in paragraph 1 or described in sub-paragraph 2 of this Part of this Schedule.

4. The following substances, namely:

---

Chorionic Gonadotrophin (HCG)
Clenbuterol
Non-human chorionic gonadotrophin
Somatotropin
Somatrem
Somatropin

---

5. Any stereoisomeric form of a substance specified or described in any of paragraphs 1 to 4 of this Part of this Schedule.

6. Any salt of a substance specified or described in any of paragraphs 1 to 5 of this Part of this Schedule.

7. Any preparation or other product containing a substance or product specified or described in any of paragraphs 1 to 6 of this Part of this Schedule, not being a preparation specified in Schedule 5.

*Appendix 5*

# Schedule 5 to the Misuse of Drugs Regulations 2001

The following is the full text of Schedule 5 as it appears in the Regulations.

Controlled Drugs Excepted from the Prohibition on Importation, Exportation and Possession and Subject to the Requirements of Regulations 24 and 26.

1. (1) Any preparation of one or more of the substances to which this paragraph applies, not being a preparation designed for administration by injection, when compounded with one or more other active or inert ingredients and containing a total of not more than 100 milligrams of the substance or substances (calculated as base) per dosage unit or with a total concentration of not more than 2.5% (calculated as base) in undivided preparations.

    (2) The substances to which this paragraph applies are acetyldihydrocodeine, codeine, dihydrocodeine, ethylmorphine, nicocodine, nicodicodine (6-nicotinoyldihydrocodeine), norcodeine, pholcodine and their respective salts.

2. Any preparation of cocaine containing not more than 0.1% of cocaine calculated as cocaine base, being a preparation compounded with one or more other active or inert ingredients in such a way that the cocaine cannot be recovered by readily applicable means or in a yield which would constitute a risk to health.

3. Any preparation of medicinal opium or of morphine containing (in either case) not more than 0.2% of morphine calculated as anhydrous morphine base, being a preparation compounded with one or more other active or inert ingredients in such a way that

the opium or, as the case may be, the morphine cannot be recovered by readily applicable means or in a yield which would constitute a risk to health.

4. Any preparation of dextropropoxyphene, being a preparation designed for oral administration, containing not more than 135 milligrams of dextropropoxyphene (calculated as base) per dosage unit or with a total concentration of not more than 2.5% (calculated as base) in undivided preparations.

5. Any preparation of difenoxin containing, per dosage unit, not more than 0.5 milligrams of difenoxin and a quantity of atropine sulfate equivalent to at least 5% of the dose of difenoxin.

6. Any preparation of diphenoxylate containing, per dosage unit, not more than 2.5 milligrams of diphenoxylate calculated as base, and a quantity of atropine sulfate equivalent to at least 1% of the dose of diphenoxylate.

7. Any preparation of propiram containing, per dosage unit, not more than 100 milligrams of propiram calculated as base and compounded with at least the same amount (by weight) of methylcellulose.

8. Any powder of ipecacuanha and opium comprising:
   10% opium, in powder,
   10% ipecacuanha root, in powder, well mixed with
   80% of any other powdered ingredient containing no controlled drug.

9. Any mixture containing one or more of the preparations specified in paragraphs 1 to 8, being a mixture of which none of the other ingredients is a controlled drug.

*Appendix 6*

# Precursor Chemicals

With the exception of those drugs which are diverted from legitimate clinical sources, many illicit drugs of abuse require the use of chemicals either to facilitate their extraction from natural products, or to form semi- or fully synthetic substances. The Convention Against Illicit Traffic in Narcotic Drugs and Psychotropic Substances (UN1988) now includes 23 such chemicals. Trade controls on these have been introduced by most member States and have been extended by some. Apart from reagents such as mineral acids and solvents that are used on a large scale primarily in cocaine processing, the essential precursor chemicals listed include those often used to manufacture amphetamine (namely 1-phenyl-2-propanone, phenylacetic acid), methylamphetamine (ephedrine and pseudoephedrine), lysergide (ergotamine, ergometrine, lysergic acid), MDMA and related drugs (safrole, isosafrole, piperonal, 3,4-methylenedioxyphenyl-2-propanone), heroin (acetic anhydride) and methaqualone (anthranilic acid, *N*-acetylanthranilic acid). These and other chemicals may be recovered from suspect shipments or from the scenes of illicit drug synthesis and submitted for laboratory analysis.

Tables A6.1 to A6.3 show the precursors and other essential reagents listed in The Controlled Drugs (Substances Useful for Manufacture) (Intra-community Trade) Regulations 1993. Category 1 chemicals are those regarded as true precursors, that is to say they form the core structure of the product drug. Category 2 chemicals are considered to be secondary precursors; they are either convertible into Category 1 precursors or are used as essential reagents. The materials in Category 3 are mostly acids and solvents, used as adjuncts in drug processing. In general terms, the legitimate industrial uses and consumption of these chemicals are least for Category 1 and greatest for Category 3.

**Table A6.1** *Category 1 chemicals*

*N*-Acetylanthranilic acid
Ephedrine
Ergometrine
Ergotamine
Isosafrole
Lysergic acid
3,4-Methylenedioxyphenylpropan-2-one
1-Phenyl-2-propanone
Piperonal
Pseudoephedrine
Safrole

Notes: (i) The legislation adds that the salts of the substances listed in this table are also covered whenever the existence of such salts is possible. (ii) Unlike in the Misuse of Drugs Act, there is no overarching control of stereoisomers, so ephedrine and pseudoephedrine are listed separately. (iii) In 2000, norephedrine was added to Table I of UN1988. In 2001, a decision was made by the UN to move acetic anhydride and potassium permanganate from Table II to Table I of UN1988. These changes have not yet been fully reflected in UK legislation.

**Table A6.2** *Category 2 chemicals*

Acetic anhydride
Anthranilic acid
Phenylacetic acid
Piperidine

**Table A6.3** *Category 3 chemicals*

Acetone
Ethyl ether
Hydrochloric acid
Methyl ethyl ketone
Potassium permanganate
Sulfuric acid
Toluene

Note: (i) The legislation adds that the salts of the substances listed in this table, except hydrochloric acid and sulfuric acid, are also covered whenever the existence of such salts is possible.

*Appendix 7*

# Relevant Stated Cases

There is a substantial body of case law concerning offences under the Act. Information is given in Appendix 8 on Appeal Court hearings which led to sentencing guidelines. The following is intended as no more than a brief index to selected cases of significance to forensic scientists. It does not include those of purely historical interest such as R-*v*-Watts, which concerned the now obsolete inclusion of dexamphetamine in the Act, or the case of R-*v*-Goodchild, a legal saga which revolved around the now defunct original definition of cannabis.

*Usability*. The currently-held view is that the principle of *de minimis* does not operate and that the proper approach to what was once called the 'usability test' is whether the prosecution can prove knowledge of possession. In other words, apart from those clearly defined situations to which Schedule 5 of the Regulations apply, the Act defines no minimum quantities below which an offence cannot be committed. This is set out in R-*v*-Boyesen, 75 Cr.App.R. 51, H.L. (1982), and superseded an earlier opposing argument that had been reached in R-*v*-Carver, 67 Cr.App.R. 352 (1978), where it had been maintained that a defendant had to be in possession of a usable quantity of a drug.

*Generic Legislation*. In considering the generic definition of phenethyl-amines, the Court of Appeal in R-*v*-Couzens and Frankel, Cr.L.R. 822 (1992), upheld the view that in paragraph 1(c) of Part I of Schedule 2 to the Act, the term 'structurally derived from' does not describe a process, but rather defines certain controlled drugs in terms of their molecular structure.

*Cannabis and Cannabis Resin*. In R-*v*-Best, 70 Cr.App.R. 21 (1979), it was held that the principle of duplicity is not compromised by a prosecution for possession of cannabis or cannabis resin. In other words it is not always necessary for one or the other to be separately specified. This is likely to be of significance only in those cases where there is insufficient material for unequivocal identification of one or the other.

*Crack Cocaine.* Even though cocaine and its salts are all treated equally as Class A drugs, in R-*v*-Russell, 94 Cr.App.R. 351 (1992), the production of crack (*i.e.* cocaine base) from cocaine hydrochloride was deemed to be an offence of production. By extension this would seem to apply to any salt–base interconversion and does not compromise the principle that if no production is alleged then the prosecution is not required to identify the form of drug present.

*Salts and Stereoisomers.* The case in R-*v*-Greensmith, 1 W.L.R. 1/24 (1983), was concerned with the specific example of cocaine, but the general point was established that the prosecution does not have to prove whether a controlled drug is in a particular stereoisomeric form or as a particular salt.

*Psilocybe Mushrooms.* In R-*v*-Stevens, Cr.L.R. 568 (1981), the Court of Appeal rejected the notion that 'preparation' had a technical pharmaceutical meaning and considered that in order for mushrooms to be prepared they had only to be treated in some way. In R-*v*-Cunliffe, Cr.L.R. 547 (1986), the example was given of deliberate drying as an act of preparation. In the subsequent case of R-*v*-Hodder, Cr.L.R. 261 (1990), concerning frozen mushrooms, the possibility was opened up that such material could be considered an 'other product' as defined in paragraph 5 of Part I of Schedule 2 to the Act.

*Appendix 8*

# Sentencing Guidelines

Maximum penalties for drugs offences are set out in the Act. Since 1982, and until recent years, sentencing in many of the larger drugs cases was based on the so-called Aramah equation. The principle was that penalties should relate to the value of the drugs seized. The street price per gram was multiplied by the weight of the seizure to get a total price. This was modified to take account of the actual purity. Thus, if the drugs were above average purity then an upward correction was made to the value. However, it had been the practice of HM Customs and Excise not to decrease the value if the purity was below the average. If the value was calculated at £100 000 or more, then a sentence of 10 years imprisonment was likely. If the value was £1 million or more, then it was 14 years. As an example, 1 kg of heroin could be worth up to £100 000. Table A8.1 shows typical current data for the common illicit drugs which might be used in an Aramah calculation.

**Table A8.1** *Typical price and purity data for use in Aramah calculations*

| Drug | Purity | Wrap Size | Price Per Gram |
| --- | --- | --- | --- |
| Amphetamine | 8% | 0.5–1.0 g | £10 |
| Cannabis – herbal | 5% | 1–4 g | £3–£4 |
| Cannabis – resin | 5% | 1–4 g | £4 |
| Cocaine | 60% | 0.2–0.4 g | £65 |
| 'Crack' cocaine | 80% | 0.1–0.2 g | £125 |
| 'Ecstasy' (MDMA *etc.*) | 80 mg/tablet | N/A | £10 (per tablet) |
| Heroin | 55% | 0.1–0.3 g | £70 |
| Lysergide (LSD) | 45 μg/unit | N/A | £3–£4 (paper square) |

Notes: (i) Prices are national averages and refer to 'street quantities', commonly termed 'wrap sizes'. (ii) Purities are typical national averages and refer to the base content of drug or THC content for cannabis products in a wide cross-section of seizures.

The Aramah equation had two major problems. Firstly, street price was subjective and often difficult to define, particularly if it was not clear on which 'street' or even in which country the drugs were to be sold. In many cases, the court had little option but to take the average value provided by the prosecution and the defence. A more fundamental objection was that if drugs become widely available, then the price would drop. The expected sentence would then also fall leading to an unacceptable situation.

The solution to these problems came about by rejecting drug valuation for the larger cases and replacing it with a more objective system based on either the weight of the pure drug or the number of dosage units. Tariffs have now been set by the Court of Appeal for the major Class A drugs: heroin, cocaine, the 'Ecstasy' group (MDMA *etc*.) and LSD as well as for the major Class B drugs: amphetamine and cannabis. Details are shown in Tables A8.2 and A8.3. Further arrangements may be necessary to deal with less common substances such as the Class A controlled drug 4-bromo-2,5-dimethoxy-α-methylphenethylamine (DOB; Bromo-STP) where tablets contain around 1 mg of the drug. Although never intended for graphical extrapolation, it will be noted from the tables that there is a non-linear relationship between sentences and weights/doses, such that smaller amounts of drug attract a disproportionately high sentence.

Because of the amounts involved, these guidelines are largely restricted to those cases involving importation, but can apply to possession and possession with intent to supply. Although no strict ruling was given, it is generally assumed that the weights of drugs shown refer to the base rather than some (arbitrary) salt form. The Court of Appeal stressed that

**Table A8.2** *Sentencing guidelines for Class A drugs*

| Drug | Weight/Dose | Sentence | Case Reference |
|---|---|---|---|
| Cocaine/Heroin | 500 g | 10 years | R-*v*-Aranguren, 16 Cr.App.R. 211 (1995) |
| Cocaine/Heroin | 5 kg | 14 years | |
| MDMA *etc*.[(i)] | 5000 tablets | 10 years | R-*v*-Warren and Beeley, 1 Cr.App.R. 120 (1996) |
| MDMA *etc*.[(i)] | 50 000 tablets | 14 years | |
| Lysergide (LSD) | 25 000 units | 10 years | R-*v*-Hurley, 2 Cr.App.R. (S) 299 (1998) |
| Lysergide (LSD) | 250 000 units | 14 years | |
| Opium | >40 kg | 14 years | R-*v*-Mashaolli, Cr.L R. 1029 (2000) |
| Opium | >4 kg | 10 years | |

[(i)]MDMA *etc*. dosage units were assumed to contain 100 mg of active drug such that 5000 tablets are equal to 500 g of pure drug. The typical content of 'Ecstasy' tablets is currently closer to 80 mg.

**Table A8.3** *Sentencing guidelines for Class B drugs*

| Drug | Weight | Sentence[i] | Case Reference |
|------|--------|-------------|----------------|
| Amphetamine | < 500 g | 2 years | R-*v*-Wijs et al., Cr.L.R. 587 (1998) |
| Amphetamine | > 500 g < 2.5 kg | 2–4 years | |
| Amphetamine | > 2.5 kg < 10 kg | 4–7 years | |
| Amphetamine | > 10 kg < 15 kg | 7–10 years | |
| Amphetamine | > 15 kg | 10–14 years | |
| Cannabis[ii] | 100 kg | 7–8 years | R-*v*-Ronchetti, Cr.L.R. 227 (1998) |
| Cannabis[ii] | 500 kg | 10 years | |

[i]The statutory maximum sentence for Class B drug offences is 14 years. [ii]Cannabis and cannabis resin should be treated equally. Unlike other drugs shown above, the weight is taken to be the seizure weight, *i.e.* no notional correction to the equivalent of pure tetrahydrocannabinol is made. For cannabis oil, 1 kg should be taken as equivalent to 10 kg of cannabis or cannabis resin. If cannabis is reclassified into Class C then the above guidelines will have to be revised.

these criteria were merely one factor in deciding appropriate sentences, and that the role of the offender, his plea and any assistance he might have given to the authorities were examples of other considerations that the court would have to consider.

## Appendix 9

# Related Legislation

Parts of the following statutes either modify some Sections of the Misuse of Drugs Act or have some relevance to other offences involving controlled drugs. Not all necessarily refer to all countries of the UK. All UK legislation since 1988 can be found at *http://www.hmso.gov.uk/ legis.htm*. A European legal database on drugs showing country profiles is available at *http://eldd.emcdda.org*.

*The Medicines Act 1968*. Regulates the licensing, manufacture and distribution of drugs in the form of medicinal products, some of which may be controlled drugs. Medicines are divided into three types: prescription only medicines can only be supplied by a pharmacist or physician; pharmacy medicines can be supplied by a pharmacist without a prescription; general sales list products can be obtained in other retail outlets.

*Road Traffic Act 1972*. Makes it an offence to be in charge of a motor vehicle while unfit to drive through drink or drugs (controlled or otherwise).

*The Customs and Excise Management Act 1979*. Extends the limited powers in the Misuse of Drugs Act against importation and exportation of controlled drugs. Provides the means for HM Customs and Excise to prosecute drug offenders involved in these activities.

*The Criminal Justice (International Cooperation) Act 1990*. One of a number of statutes created to discharge UK responsibilities to UN1988. Regulates the licensing, manufacture and distribution of substances (precursors) useful for the production of illicit drugs. See Appendix 6 for a list of precursor chemicals.

*The Drug Trafficking Act 1994*. Enables the UK to meet further obligations under UN1988. It replaced the Drug Trafficking Offences Act 1986. Following a conviction, a court is able to confiscate the assets of those who have benefited from drug trafficking. The Act applies to England and Wales only.

*The Crime and Disorder Act 1998.* Created Drug Testing and Treatment Orders later consolidated by the Powers of Criminal Courts (Sentencing) Act 2000. These orders are designed as a community sentence for those aged 16 or above who commit crime to fund a drug habit.

*The Criminal Justice and Police Act 2001.* Section 38 amends Section 8(d) of the Misuse of Drugs Act 1971 to read 'administering or using a controlled drug which is unlawfully in any person's possession at or immediately before the time when it is administered or used'. Section 8(d) was previously only concerned with 'smoking cannabis, cannabis resin or prepared opium'.

# Appendix 10

# The Origin and Main Characteristics of the Major Drugs of Abuse

### 10.1  Amphetamine and Methylamphetamine

The world-wide production and consumption of amphetamine and methylamphetamine shows clear geographical trends. In Europe, amphetamine is much more common than methylamphetamine, but in North America and the Far East this situation is reversed. Despite global controls on the trade in precursor chemicals, most illicit amphetamine is manufactured *via* the Leuckart route from 1-phenyl-2-propanone (P2P, BMK) together with formamide or ammonium formate. This and other syntheses from P2P produce a racemic product often seen as a white or off-white powder and less commonly as tablets. Caffeine is added to amphetamine at source, but glucose and other sugars are used as subsequent cutting agents. Methylamphetamine is generally synthesised by the stereoselective reduction of either L-ephedrine or D-pseudoephedrine. It may occur as powders or tablets, but pure crystalline D-methylamphetamine hydrochloride ('Ice') is rarely seen outside the Far East.

### 10.2  Cannabis

Herbal cannabis consists of the dried flowering tops and leaves of plants of the genus *Cannabis*. Cannabis resin is a compressed solid made from the resinous parts of the plant. Herbal cannabis imported into Europe may originate from West Africa, the Caribbean or South East Asia, but cannabis resin derives largely from either North Africa or Afghanistan. Cannabis resin is usually produced in 250 g blocks (so-called 9 ounce bars), many of which carry a brandmark impression. Improved seed varieties and procedures such as artificial heating and lighting,

cultivation in nutrient solutions and propagation of cuttings
plants lead to a high production of flowering material
mes known as 'skunk'), where the THC content may be in excess
of 20%. In the UK and some other EU states, there is licensed cultivation
of cannabis for the production of hemp fibre, but the THC content of
these plants is less than 0.3%. The average 'reefer' cigarette contains
around 200 mg of herbal cannabis or cannabis resin.

### 10.3    Cocaine

The shrub *Erythroxylon coca* is cultivated widely on the Andean ridge in
South America and is the only known natural source of cocaine.
The leaves, which contain up to 1% cocaine, are extracted with solvents
and other reagents to produce cocaine hydrochloride. Potassium
permanganate is used in the process to destroy cinnamoylcocaine;
without this the final product will be of a low quality, discoloured and
difficult to crystallise. Cocaine hydrochloride is a white crystalline
powder. Typical cutting agents are lactose and other sugars, lidocaine
and similar local anaesthetic drugs, caffeine, paracetamol, phenacetin
and dimethyl terephthalate.

### 10.4    Diamorphine (Heroin)

Diamorphine, the primary constituent of heroin, is produced by the
acetylation of crude morphine obtained from opium. Until the late
1970s, nearly all heroin consumed in Europe came from South East Asia,
but now most originates from South West Asia, an area centred on
Pakistan, Afghanistan and Turkey. Typically seen as a brown powder, it
has variable amounts of other opium alkaloids (*e.g.* monoacetylmorphine, noscapine, papaverine and acetylcodeine) as well as adulterants
such as caffeine and paracetamol. It is believed that the latter are added
to heroin either at the time of manufacture or during transit. Other less
common psychoactive adulterants include phenobarbitone, methaqualone and diazepam. Brown heroin may be 'smoked' by heating the solid
on a metal foil above a small flame and inhaling the vapour. Those
intending to inject brown heroin must first solubilise it with, for example,
citric acid or ascorbic acid.

### 10.5    Lysergide (LSD)

LSD is the diethylamide of lysergic acid. It is synthesised from
ergotamine produced by microbial fermentation. It is generally believed

that most of this semi-synthetic drug is now produced in the US, but the preparation of dosage units by dipping or spotting paper squares is more widespread. These dosage units usually bear coloured designs featuring cartoon characters, geometric and abstract motifs. The only illicit analogue of LSD to have appeared is the *N*-methyl-*N*-propyl isomer of lysergide known as LAMPA.

## 10.6   MDMA

First synthesised by the Merck Company in 1914, MethyleneDioxy-MethylAmphetamine (MDMA) was never marketed as a medicinal product. In the UK, large-scale consumption of illicit material did not begin until the late 1980s. Clandestine production is still largely centred in Europe. A number of homologous compounds with broadly similar effects, *e.g.* MDA (MethyleneDioxyAmphetamine), MDEA (MethyleneDioxyEthylAmphetamine) and MBDB [*N*-**M**ethyl-1-(1,3-**B**enzo-**D**ioxol-5-yl)-2-**B**utanamine] have appeared, but have proved less popular. These substances are collectively known as the 'Ecstasy' drugs. They invariably occur as tablets, which carry a characteristic impression (logo). Apart from the active drug, tablets contain a bulking agent such as lactose and smaller quantities of binders. Mixtures of controlled drugs in illicit tablets are not common and the presence of dangerous 'contaminants' is almost unknown.

Some of the chemical properties of these drugs are shown in Table A10.1.

**Table A10.1** *Selected chemical properties of the major drugs of abuse*

| Name | Molecular Weight of Base (Daltons) | Typical Salt | Base Content of Salt |
|------|------------------------------------|--------------|----------------------|
| Amphetamine | 135.2 | Sulfate | 73% |
| Cocaine | 303.4 | Hydrochloride | 89% |
| Diamorphine | 369.4 | Hydrochloride | 91% |
| Lysergide (LSD) | 323.4 | Tartrate | 78% |
| MDMA | 193.2 | Hydrochloride | 84% |
| Methylamphetamine | 149.2 | Hydrochloride | 80% |

*Appendix 11*

# Drugs and the Internet

As with many other topics, the Internet has grown to become a major source of information on drugs. Unlike much printed literature, there is little quality control and information can range from the reliable to that which is anecdotal, inaccurate or deliberately misleading. Some web-sites can be far more ephemeral than printed matter. Table A11.1 provides a list of the official sites of the principal agencies involved in drug control, public policy or related areas.

**Table A11.1** *Web-sites for the major agencies concerned with drug control, public policy or related areas*

| *Address* | *Host Organisation* |
| --- | --- |
| *eldd.emcdda.org* | A European legal database on drugs |
| *www.drugabuse.gov* | National Institute on Drug Abuse (NIDA) |
| *www.drugscope.org.uk* | Drugscope |
| *www.emcdda.org* | European Monitoring Centre for Drugs and Drug Addiction (EMCDDA) |
| *www.europol.eu/home.htm* | Europol |
| *www.hmce.gov.uk* | HM Customs and Excise |
| *www.hmso.gov.uk/legis.htm* | All UK legislation since 1988 |
| *www.homeoffice.gov.uk* | Home Office |
| *www.incb.org* | UN International Narcotics Control Board |
| *www.interpol.int* | Interpol (ICPO) |
| *www.open.gov.uk/mca/homemain.htm* | Medicines Control Agency |
| *www.undcp.org* | UN Drug Control Programme |
| *www.usdoj.gov/dea* | Drug Enforcement Administration (USA) |
| *www.whitehousedrugpolicy.gov* | Office of National Drug Control Policy (USA) |

# Subject Index

Separate indexes to all controlled drugs can be found in Chapter 2 (pp. 7–17)

Advisory Council on the Misuse of
Drugs, 3, 4, 84
alkaloid, 18, 47, 71, 79, 90, 108
alkyl nitrite, 84
amphetamine, 1–3, 5, 23, 30, 44–46,
51, 60, 61, 65, 69, 71, 73, 74, 83,
107
anabolic/androgenic steroid, 19–21,
29, 36, 37, 67, 68, 75
amfepramone, *see* diethylpropion
Aramah calculation, 102, 103
Ayahuasca, 47


BAN, *see* British Approved Name
barbiturate, 2, 19, 29, 35, 36
benzodiazepine, 29, 67, 71, 82
benzoylecgonine, 27
*N*-benzylpiperazine, 63
British Approved Name, 38
bromo-STP, *see* 4-bromo-$\beta$,2,5-tri-
methoxyphenethylamine
4-bromo-$\beta$,2,5-trimeth-
oxyphenethylamine, 31, 40, 42,
45, 70, 74, 103


cannabinoid, 26, 82
cannabinol and derivatives, 20, 25,
26, 41, 43, 63, 64, 76, 77, 84, 90,
104

cannabis, 1–3, 5, 25, 26, 66, 75–77,
82, 84, 90, 100, 102, 104, 107,
108
cannabis cigarette, 108
cannabis seed, 66, 82, 107
$\beta$-carboline, 46, 47, 50
cases
R-*v*-Aranguren, 103
R-*v*-Best, 100
R-*v*-Boyesen, 100
R-*v*-Carter, 76
R-*v*-Carver, 100
R-*v*-Couzens and Frankel, 43, 100
R-*v*-Cunliffe, 101
R-*v*-Farmer, 83
R-*v*-Goodchild, 100
R-*v*-Greensmith, 101
R-*v*-Hodder, 101
R-*v*-Hurley, 103
R-*v*-Karagozlu, 69
R-*v*-Mashaolli, 103
R-*v*-Ronchetti, 104
R-*v*-Russell 67, 101
R-*v*-Stevens, 70, 101
R-*v*-Warren and Beeley, 103
R-*v*-Watts, 100
R-*v*-Wijs, 104
cathine, 23, 83
cathinone, 60, 83
CND, *see* United Nations

# RSC Paperbacks

RSC Paperbacks are a series of inexpensive texts suitable for teachers and students and give a clear, readable introduction to selected topics in chemistry. They should also appeal to the general chemist. For further information on all available titles contact:

Sales and Customer Care Department, Royal Society of Chemistry,
Thomas Graham House, Science Park, Milton Road, Cambridge CB4 0WF, UK
Telephone: +44 (0)1223 432360; Fax: +44 (0)1223 423429; E-mail: sales@rsc.org

Recent Titles Available

**The Chemistry of Fragrances**
*compiled by David Pybus and Charles Sell*
**Polymers and the Environment**
*by Gerald Scott*
**Brewing**
*By Ian S. Hornsey*
**The Chemistry of Fireworks**
*by Michael S. Russell*
**Water (Second Edition): A Matrix of Life**
*by Felix Franks*
**The Science of Chocolate**
*by Stephen T. Beckett*
**The Science of Suger Confectionery**
*by W.P. Edwards*
**Colour Chemistry**
*by R.M. Christie*
**Understanding Batteries**
*by Ronald M. Dell and David A.J. Rand*
**Principles of Thermal Analysis and Calorimetry**
*Edited by P.J. Haines*
**Green Chemistry: An Introductory Text**
*by Mike Lancaster*
**The Misuse of Drugs Act: A Guide for Forensic Scientists**
*by L.A. King*

Future titles may be obtained immediately on publication by placing a standing order for RSC Paperbacks. Information on this is available from the address above.

# THE MISUSE OF DRUGS ACT
## A Guide for Forensic Scientists